Urban Sustainability

The Urban Sustainability Book Series is a valuable resource for sustainability and urban-related education and research. It offers an inter-disciplinary platform covering all four areas of practice, policy, education, research, and their nexus. The publications in this series are related to critical areas of sustainability, urban studies, planning, and urban geography.

This book series aims to put together cutting-edge research findings linked to the overarching field of urban sustainability. The scope and nature of the topic are broad and interdisciplinary and bring together various associated disciplines from sustainable development, environmental sciences, urbanism, etc. With many advanced research findings in the field, there is a need to put together various discussions and contributions on specific sustainability fields, covering a good range of topics on sustainable development, sustainable urbanism, and urban sustainability. Despite the broad range of issues, we note the importance of practical and policy-oriented directions, extending the literature and directions and pathways towards achieving urban sustainability.

The series will appeal to urbanists, geographers, planners, engineers, architects, governmental authorities, policymakers, researchers of all levels, and to all of those interested in a wide-ranging overview of urban sustainability and its associated fields. The series includes monographs and edited volumes, covering a range of topics under the urban sustainability topic, which can also be used for teaching materials.

Jiayu Wu

Urban Green Space Equity in China

Nation, City, and Individual

Springer

Jiayu Wu
Institute of Landscape Architecture
Zhejiang University
Hangzhou, Zhejiang, China

ISSN 2731-6483 ISSN 2731-6491 (electronic)
Urban Sustainability
ISBN 978-981-95-5770-7 ISBN 978-981-95-5771-4 (eBook)
https://doi.org/10.1007/978-981-95-5771-4

This work was supported by National Science Foundation of China (32271935, 32471932, 51908488).

This Springer imprint is published by the registered company Springer Nature Singapore Pte Ltd.
The registered company address is: 152 Beach Road, #21-01/04 Gateway East, Singapore 189721, Singapore

Preface

Urban green spaces (UGS) support the integrity of urban ecosystems by enabling pollution control, noise reduction, and temperature regulation. They also contribute to enhancing urban residents' quality of life and promoting public health in cities. However, extensive research across various scales has revealed that the distribution of UGS is not always equitable. UGS inequity is widely acknowledged as one of the most critical environmental justice issues.

Research on UGS equity spans multiple scales: individual (micro), urban (meso), and national/global (macro). The dimensions of green equity encompass procedural justice, distributional justice, and interactional justice. Within policy frameworks for equity, regulatory, administrative, planning, and implementation systems exhibit varying degrees of inequity. Socioeconomic factors—such as income, age, gender, and ethnicity—and their correlation with disparities in UGS access are extensively studied. Multifaceted dimensions of UGS, including availability, visibility, accessibility, and aesthetic quality, have been proposed, with diverse analytical methods and cutting-edge technologies applied to advance equity research.

Chapter 1 of this book reviews the conceptual frameworks and recent progress. Chapters 2–4 introduce perspectives on UGS equity at national, urban, and individual scales, respectively, supported by empirical case studies that highlight cutting-edge methodologies and insights at each scale. Finally, Chap. 5 discusses future trends in UGS equity research.

Hangzhou, Zhejiang, China — Jiayu Wu

Competing Interests The author has no competing interests to declare that are relevant to the content of this manuscript.

Contents

List of Abbreviations

ABE	Anchor-based green exposure
AOI	Area of interest
API	Application programming interface
GBDT	Gradient boosting decision tree
GEE	Google Earth Engine
GINI	Gini Index
GVI	Green View Index
MBE	Mobility-based green exposure
NDVI	Normalized Difference Vegetation Index
NEAP	Neighborhood effect averaging problem
NEPP	Neighborhood effect polarization problem
PGS	Public green space
POI	Point of interest
RBE	Residence-based green exposure
SHAP	Shapley additive explanations
UGS	Urban green space

Chapter 1
Overview of Research Progress on Urban Green Space Equity

1.1 Introduction

A substantial body of empirical research demonstrates that urban green space (UGS) delivers a wide range of co-benefits that can be grouped into ecological, health, and social domains. Ecologically, as a key subsystem of the urban ecosystem, green spaces sequester carbon and release oxygen; conserve water; regulate microclimates; reduce noise and particulates; mitigate pollution; and sustain biodiversity (Gunawardena et al. 2017). In doing so, they support natural processes such as the water cycle, gas exchange, and nutrient cycling, and they provide platforms for ecological education and research (Knobel et al. 2021). On the health front, UGS is associated with lower risks of all-cause and cardiorespiratory mortality; it helps buffer heat waves and the urban heat island; improves respiratory health; and offers venues for physical activity and psychological restoration, alleviating stress and depression (Diener and Mudu 2021). By attenuating flood and fire risks, storing and detaining stormwater, and offering emergency refuge, green infrastructure also strengthens urban disaster prevention, mitigation, and resilience. Socially, green space functions as a public good with positive externalities: it enhances urban image and attractiveness, stimulates tourism and related industries, carries historical and cultural meanings and landscape aesthetics, and supplies high-quality spaces for leisure and recreation. Equitably accessible green public space can reinforce social cohesion and inclusion, advancing environmental justice and social harmony.

However, as public social resources, UGS exhibit widespread inequities across scales, with significant disparities in access driven by income levels, age groups, racial backgrounds, and social identities (Chen et al. 2021; Lu et al. 2024; Wu et al. 2019). Over the past two decades, the international academic community has conducted extensive theoretical and practical research on this issue, recognizing inequitable UGS accessibility as one of the most critical environmental justice concerns.

J. Wu, *Urban Green Space Equity in China*, Urban Sustainability,
https://doi.org/10.1007/978-981-95-5771-4_1

Against this backdrop, a dedicated focus on China is both warranted and timely. First, China is among the countries with the largest scale and fastest pace of urbanization, alongside accelerated population aging. As a tool for integrated "low-carbon–health–resilience" governance, the equitable allocation of green space directly affects heat-risk exposure, the burden of chronic disease, and intergenerational well-being. Recent studies have documented substantial inequalities in the social distribution of heat exposure and green space services in China's major cities. Second, China is advancing ecological civilization and piloting policies such as the "park city" model and the "15-minute living circle," creating a distinctive institutional window for embedding distributional, procedural, and interactional justice into territorial spatial planning and urban regeneration. Related evidence indicates that an elderly-oriented 15-minute approach and high-quality streetscape greening can markedly improve equity at the neighborhood scale. Third, China possesses world-leading data and methodological conditions: high-density street-view imagery, navigation and road network data, and mobile-signaling records—among other multi-source datasets—enable an end-to-end assessment spanning nominal supply, actual exposure, and experienced quality. Fourth, China's pronounced regional gradients and diverse urban-size hierarchy offer a natural comparative arena for testing mechanisms of spatial inequality—North–South/East–West and center–periphery—and for generating transferable evidence and toolkits for urban agglomerations across the Global South. In sum, deepening research on green space equity in China can meet domestic needs for health and climate-adaptation governance while, through large samples, multi-source data, and policy practice, enriching and advancing international frontiers in environmental justice and sustainable development.

1.2 From Environmental Justice to Green Equity

Environmental justice is the principle that all people, regardless of race, ethnicity, or socioeconomic status, have the right to a safe and healthy environment. Everyone should have equal opportunities to access environmental benefits such as clean air, water, and soil, and to be protected from environmental hazards such as pollution and hazardous waste (Schweizer 1999). The roots of this principle can be traced to the sixth century under Emperor Justinian of the Roman Empire. His codification of law granted the public rights to the commons, asserting that the air, water, fish, the sea, and the coastline are common property freely available for everyone's use (Emmett 2015). This idea is regarded as an early precursor of environmental equity and helped lay the groundwork for later environmental protection movements. In modern times, the concept took shape and gained recognition through environmental equity movements worldwide (Mitchell et al. 2015).

The environmental justice movement began in the United States in the late 1960s, led by communities of color. These communities were often impoverished and disproportionately burdened by environmental pollution and health risks associated with industrialization and urbanization. Residents organized to claim their environmental rights and demand justice, and some groups allied with broader social movements, including the civil rights and antiwar movements (Mohai and Bryant 1992). One emblematic episode was the Memphis sanitation strike, in which Black sanitation workers facing harsh working conditions and low wages mounted a 2-month strike. The strike ultimately succeeded and became a landmark moment at the intersection of environmental equity and the civil rights movement (Estes 2000).

In the early 1970s, the Love Canal disaster in New York State became a milestone in the environmental justice movement. From 1943 to 1953, Hooker Chemical was authorized to use the canal as a landfill. It disposed of roughly 20,000 tons of chemical waste sealed in drums and buried, then capped the site with soil. The parcel was later sold to the Niagara Falls Board of Education, which built the 99th Street School on the site. Local authorities then redeveloped the area as housing for low-income and single-person households. Construction at Love Canal triggered the release of buried toxic chemicals. During rainfall, contaminants mixed with stormwater to create hazardous surface runoff, and residents suffered a range of illnesses, including miscarriages, infant mortality, congenital malformations, epilepsy, and rectal bleeding. Public outrage led residents and environmental organizations to undertake years of protest and litigation, drawing national attention. In 1978, President Jimmy Carter declared a federal public health emergency, called for the allocation of federal funds, and directed federal disaster-assistance agencies to help the City of Niagara Falls remediate the site (Blum 2008). The episode had far-reaching implications for subsequent legislation, environmental policymaking, and urban planning, heightening public awareness of environmental inequities. Love Canal became a symbol of the environmental justice movement in the United States and a reminder to governments and corporations to prioritize environmental fairness, particularly for vulnerable groups.

Research emerging from the environmental justice movement has revealed the inherent spatial inequalities of urban environments (Bryant 1995), documented the adverse impacts of unevenly distributed environmental hazards on low-income residents and communities of color, and substantially advanced practitioners' understanding of urban environmental conditions. Nevertheless, the field's theoretical vantage point has often remained underdeveloped constrained by limited attention to the historical processes that produce inequality and by a persistent focus on hazards as the primary object of inquiry.

As inquiry into environmental injustice deepened, the definition of "environment" broadened as well. Activists criticized the narrow conception that equated environment with wilderness or pristine nature and advanced a more expansive understanding—namely, the places where we live, work, and play (Novotny 2000). The significance of this shift is hard to overstate: it foregrounds the intrinsic connections and interactions between people and their surroundings, enabling research and practice to consider human–environment relations more holistically (Schlosberg 2013). It also recognizes a basic reality: beyond pollution burdens, the material conditions of the environments in which the poor live are themselves deprived (Laituri and Kirby 1994). Accordingly, scholars and advocates widened the agenda of environmental justice to encompass unequal protection and the distribution of environmental goods and services, along with related domains. Limited access to environmental benefits—or to nature's services—can likewise be understood as environmental inequality (Opotow and Clayton 1994).

In October 1991, the United States hosted the first national summit of environmental leaders from communities of color. The summit drafted and adopted seventeen principles of environmental justice, formally establishing the concept and framework of environmental justice (Boone et al. 2009). One principle affirmed the necessity of coordinated ecological policies for both urban and rural areas, calling for cleanup and renewal in balance with nature and for equitable opportunities for all to access the full range of resources. The summit catalyzed a new understanding of environmental justice, highlighted the importance of green, ecological, and sustainable development, and opened fresh perspectives and dimensions for research in the field.

The Journal of Social Issues published a special issue on green equity in 1994 (*Green Justice: Conceptions of Fairness and the Natural World*), marking the first time that the unequal distribution of natural resources was explicitly framed under this theme (Opotow and Clayton 1994). The issue introduced foundational questions linking fairness with the natural environment and argued that urban areas are the primary arena in which green space equity problems arise (Laituri and Kirby 1994), with urbanization generating widespread challenges of this kind. The editors noted that much work remained to be done, especially the need to foreground the environmental perspectives of marginalized groups. They called for comparative studies of access to natural resources among groups occupying central versus peripheral positions, as well as across differences in age, gender, income, and cultural traditions (Opotow and Clayton 1994). The publication of this special issue laid a solid foundation for the subsequent development of green space equity research, deepened reflection on the relationship between social justice and the natural environment, and charted a consequential direction for future inquiry.

In the early 1990s, research on green equity focused primarily on assessing regional equality, with scholars commonly using per capita park green space indicators as the principal metric. This line of work revealed the uneven distribution of park benefits across urban populations (Wolch et al. 2005). In the United States in particular, studies showed that park provision disproportionately favored White and affluent residents, while communities of color faced greater constraints on accessibility and opportunities for use.

At that time, scholarly debate over differences in green space use centered on three theoretical frameworks: the marginality hypothesis, the ethnicity hypothesis, and the acculturation or assimilation hypothesis (Krymkowski et al. 2014). The marginality hypothesis posits that people of color face socioeconomic barriers—travel costs, equipment expenses, park entrance fees, and limited leisure time—that restrict their park use (Edwards 1981; Stoep 1993). The ethnicity hypothesis emphasizes distinctive subcultural styles among different ethnic groups that shape leisure preferences and activity choices (Johnson et al. 1998; Washburne 1978). The acculturation or assimilation hypothesis argues that the park-use patterns of people of color differ from those of White populations because they have not fully adapted to, or adopted, mainstream values (Chick 2005; Stodolska 2005; Stodolska and Jackson 1998). These frameworks, however, have notable shortcomings. First, they tend to embody entrenched Anglocentrism, conflate race and ethnicity, overstate the inevitability and legitimacy of assimilation, and, crucially, neglect the role of space and place in shaping park-use choices. Second, the data used to test these hypotheses are seldom representative, which severely limits the generalizability of findings. Third, the theories have not been comprehensively evaluated; for example, tests of marginality rarely benchmark against appropriate demographic controls, the discrimination hypothesis has been infrequently examined, and assessments of the subcultural account often rely on crude indicators of group membership rather than validated cultural measures (Krymkowski et al. 2014).

Research on public policymaking for green equity has typically rested on a basic assumption: increasing the supply of green space and providing convenient transportation will satisfy public demand (Yan et al. 2024). By the late 1990s, geographers, planners, and psychologists began to offer alternative perspectives that highlight the decisive roles of cognitive, social, and socio-spatial factors in shaping park-use patterns. The socio-spatial explanatory framework proposes that park use results from multiple interacting factors, including spatial proximity, accessibility barriers, facility appeal, landscape attributes, safety and security, maintenance levels, the availability of substitute recreational opportunities, and users' perceptions and cognition (Burgess et al. 1988; Lee et al. 2005; Smith 1980). This turn made green equity research more complex, as social stratification and cultural diversification generated heterogeneous needs and growing differentiation among social groups. Identifying the needs of different groups and providing more precisely targeted services has since become a central research priority (Yan et al. 2024).

In the first decade of the twenty-first century, research on green equity underwent a major paradigm shift. With the widespread adoption of geographic information systems and a deepening understanding of environmental justice, the focus moved from simple regional comparisons to more complex analyses of spatial distributive equity, and methods became increasingly refined (Rigolon 2016). Studies in this period continued to examine park proximity and the equity of accessible park area. Traditional accessibility metrics that relied on simple distance were broadened by a comprehensive framework for measuring facility access that encompassed five categories: container, coverage, minimum distance, travel cost, and gravity (Talen 2003). This framework exerted a profound influence on subsequent work. Researchers began to apply more sophisticated spatial analyses that considered not only distance but also population density, socioeconomic characteristics, and other dimensions (Wolch et al. 2005). The diversification of methods and indicators produced a number of notable findings. A hallmark study in a large U.S. metropolitan region, for example, identified a complex pattern: Black residents were more likely than White residents to live within walking distance of parks, yet White residents had access to a larger total park area, resulting in more severe crowding at parks serving Black communities. Similar distributional patterns were documented across multiple cities, particularly in major metropolitan areas. This era of research also showed how evolving urbanization patterns affected park equity. The traditional dichotomy of minority inner cities versus White suburbs began to erode as poverty increasingly suburbanized (Kneebone and Garr 2010). Scholars further demonstrated that differences in park use arose not only from geographic accessibility but also from sociocultural dynamics; some parks were perceived as unsafe or as spaces that "belong" to other groups (Byrne 2012).

From 2010 to 2014, the field expanded beyond single measures of distance or area to multi-perspective quality assessment, yielding a more systematic framework for park-related environmental equity. Evidence from large metropolitan regions showed that low-income communities of color not only had less park space per capita but also faced poorer facility quality and greater crowding, echoing earlier distributional findings (Sister et al. 2010). Studies in medium-sized cities documented significant gaps in facility counts, maintenance standards, and safety in disadvantaged neighborhoods. Researchers also incorporated natural attributes such as tree-canopy cover, with comparative analyses indicating that parks in higher-income areas often possess substantially more shade than those in lower-income areas (Vaughan et al. 2013).

In 2014, the article *Urban green space, public health, and environmental justice: The challenge of making cities "just green enough"* made a major impact (Wolch et al. 2014). It successfully integrated urban green space, environmental justice, and public health. By systematically tracing research on green space accessibility and social equity, the article proposed a "just green enough" strategy to reconcile greening with gentrification. This approach emphasizes not only ecological restoration but also the planning of green space according to the specific needs of local communities, in order to avoid the intensified inequalities that can accompany large-scale redevelopment. Through small-scale interventions and robust community participation, it advocated a more inclusive and sustainable model for green space development. From that point on, publications on green equity grew rapidly. The article also presented a detailed green equity analysis in a Chinese city, Hangzhou, which helped catalyze a wave of related studies by Chinese scholars.

From 2014 to 2022, research on green equity expanded markedly, due in large part to the active contributions of Chinese scholars. Through sustained empirical work and theoretical synthesis, the field reached broad consensus on core conceptual frames, evaluation methodologies, and research paradigms for park-related equity (Yan et al. 2024). Studies typically assessed multiple facets of urban parks—such as quantity, quality, and type—and linked these with the sociodemographic composition of residents within spatial units to evaluate differences in accessibility across social groups and, in turn, the fairness of resource allocation. Findings from this phase can be summarized along two main lines. First, mainstream research confirmed pronounced spatial unevenness in the distribution of parks: older adults, children, migrant populations, and ethnic minorities generally face lower levels of accessibility (Chen et al. 2021; Kim et al. 2023; Wu et al. 2022; Zhang et al. 2022). Second, areas with higher housing prices tend to enjoy higher-quality park resources (Chen and Yue 2020; Fan et al. 2017), and, relative to suburban areas, central districts often show greater accessibility to parks (Chen and Yue 2020; Tian et al. 2021).

In recent years, scholars have recognized both the global and regional dimensions of green equity, extending analyses beyond domestic cities and neighborhoods to international comparisons (Chen et al. 2022b). Work has examined imbalances in resources and environmental benefits between the Global North and South, and the disadvantages faced by low-income and developing countries in environmental governance. Green equity is now understood to involve not only the distribution of resources but also inequities in environmental access, participation, and governance (Rigolon 2016; Wang et al. 2022, 2024). Comparative studies across lines of ethnicity, gender, age, and class have gained momentum; for example, evidence shows that women and racial or ethnic minorities are often more exposed to environmental burdens than men and White populations, underscoring persistent gendered and racialized inequalities (Wu et al. 2022).

The field continues to evolve, encompassing climate change, urban green infrastructure, and public and environmental health. Researchers increasingly emphasize the complexity of green equity, noting that it is shaped by political, economic, cultural, and institutional factors and thus requires interdisciplinary inquiry that tightly links environmental issues with social, political, and cultural questions. At the same time, scholars are exploring new solutions, including community engagement and sustainability-oriented approaches. Overall, green equity is a complex and enduring challenge whose scope is steadily widening. Future research must probe its multiple dimensions and scales more deeply and comprehensively, while developing more effective pathways to solutions.

1.3 Research Scales in UGS Equity Studies

As the field has advanced, scholars have turned to how green equity manifests across different geographic scales and sociocultural contexts (Chen et al. 2022b; Xu et al. 2018). Countries and regions vary markedly in green space planning and management, and the uneven distribution of urban green space displays complex, scale-dependent patterns for three main reasons.

First, both the social differentiation of green equity and the indicators used to assess it differ by unit of analysis. At the national scale, research often examines how green space in the built environment relates to levels of urban economic development and industrial structure. At the city scale, studies typically focus on residential neighborhoods and communities, evaluating spatial inequalities in green space accessibility based on area-level socioeconomic attributes such as median income, housing prices, employment rates, and educational attainment. At the individual scale, work drills down to human activity, assessing how income, gender, age, and race shape exposure to greenery (Kronenberg et al. 2020; Wu et al. 2019).

Second, the modifiable areal unit problem—MAUP—introduces strong scale dependence into evaluations of green equity (Chen et al. 2022a). Changing the scale can alter the strength of associations between socio-demographic and environmental variables, as well as the spatial patterning of inequality and its dominant drivers

(Fernández and Wu 2016). For example, economic development may be a primary driver of green inequality at coarse scales, yet at finer scales its influence may be displaced or attenuated by other factors. This scale dependence can cause policy interventions designed for one scale to fail at another.

Finally, the types and functional characteristics of green space differ across scales. From natural-area green space—ecological corridors, nature reserves—to built-environment green space—urban parks, community greens—and down to zones and facilities within parks, each scale carries distinct ecological functions and social values (Kronenberg et al. 2020). A multi-scale analytical framework is therefore essential to fully capture distributional features and fairness concerns.

Accordingly, building a multi-scale evaluation system for green equity has significant theoretical and practical value. Such a system should: (1) Compile a multi-scale database of green space distributions that spans natural areas and the built environment; (2) Develop diagnostic indicator systems suited to different scales; (3) Compare how social differentiation relates to green equity at each scale; (4) Reveal the spatial heterogeneity and scale diversity of green equity challenges. A systematic, multi-scale approach will not only deepen theoretical understanding of green equity but also provide targeted scientific evidence for policymaking across administrative levels.

1. National-Scale Green Equity Assessment

Research on green equity at the national confronts more complex and multidimensional challenges. This requires addressing both broad geographical, economic, social, and cultural influences, as well as transboundary disparities in the physical distribution of UGS, population exposure patterns, and equity dynamics (Lu et al. 2024; Song et al. 2021; Wu and Kim 2021). At these scales, UGS equity assessment entails statistical analyses of nationwide or global green resources, encompassing total UGS area, spatial configuration, typological diversity, and regional green coverage. Simultaneously, it must consider impacts from demographic distributions, economic development disparities, urbanization gradients, global climate change effects on UGS, and transboundary conservation governance. Such assessment facilitates international cooperation in UGS management while contributing substantively to Sustainable Development Goals (SDGs), particularly SDG 11.

Notably, macro-scale assessments demand heightened attention to individual mobility and spatial heterogeneity in UGS environments. Accelerated globalization and frequent cross-border migration amplify mobility's equity impacts, while significant variations in UGS quality and accessibility across regions compound disparities. Consequently, these studies necessitate geospatial breadth spanning transnational boundaries, extended temporal depth capturing decadal dynamics, emerging technologies integrating multi-source satellite data, population mobility datasets, and AI-driven analytics. Subsequent analyses will examine spatial heterogeneity and temporal dynamics of UGS distribution inequalities at national level.

2. Urban-Scale Green Equity Assessment

Urban-scale green space equity assessment expands to meso-scale units such as residential districts, neighborhoods, and administrative zones (Wolch et al. 2014;

Wu et al. 2020b), where accessibility measurement models are widely applied. The container model, as the earliest approach, artificially divides study areas into subunits (e.g., administrative zones) for analysis, offering operational simplicity. Subsequently, the coverage model improved upon administrative boundary limitations by incorporating park service radii based on green space categories and sizes, combined with park service area methods and Voronoi polygon partitioning—though still presenting crude behavioral representations. The gravity-based model introduced park attractiveness and friction factors (distance, travel cost) that impede access, yet remained deficient in accounting for UGS quality and multimodal transportation options. The floating catchment model subsequently enabled accessibility estimation across multiple travel modes (driving, public transit, cycling, walking), differentiated distance thresholds for varying UGS sizes, and explicitly considered supply–demand interactions between residents and green spaces. This framework spawned derivative sub-models including the Two-Step Floating Catchment Area (2SFCA), Population-Weighted Distance (PWD), and Variable-width Floating Catchment Area (VFCA). The Comprehensive Floating Catchment Area (CFCA) model further incorporated transport network constraints and distance-decay effects of UGS attractiveness, highlighting street greening's contributions to UGS appeal and green inequity mitigation.

Over recent decades, UGS assessment models have evolved along two dominant trajectories: First, methodological refinement manifests through increasingly precise service area delineation, diversified transportation mode integration, and sophisticated supply–demand relationship modeling (Dony et al. 2015). Distance measurement has progressed from straight-line estimates to network path simulations and real-time navigation routing, significantly enhancing accuracy. Second, persistent neglect of UGS heterogeneity not only impedes accurate assessment of quality-based accessibility reflecting comprehensive user experiences but also risks overestimating or underestimating inequity levels. Consequently, improvement strategies remain confined to qualitative policy discussions. Following sections will therefore examine differential outcomes in urban green space equity when considering: open community green spaces, street greening integration, and multidimensional UGS service capacity evaluations (LaCour-Little 2001; Wu et al. 2020a).

3. Individual-Scale Green Equity Measurement

Individual-scale green equity typically encompasses three measurement dimensions of green exposure: availability, visibility, and accessibility (Kronenberg et al. 2020). Availability refers to an individual's capacity to access nearby green spaces, commonly quantified through metrics such as greening rate, vegetation coverage, and the Normalized Difference Vegetation Index (NDVI). Visibility specifically measures the proportion of visible vegetation in streetscapes, primarily assessed via the green view index (Yan et al. 2024). Accessibility measurement models have evolved from foundational gravity models to include container model, coverage model, gravity-based model, and floating catchment model, simulating supply–demand relationships between residents and green spaces. Recent advances in environmental behavior research recognize that individual mobility patterns significantly influence exposure assessments. Beyond static geographical contexts like residential or workplace locations, daily movement trajectories must now be considered. The failure to account for individual mobility and spatial heterogeneity in urban green spaces—due to challenges such as the Uncertain Geographic Context Problem (UGCoP) and daily mobility bias—compromises measurement accuracy of actual environmental exposure, potentially leading to under- or over-estimation of green equity. Consequently, Chap. 2 will systematically characterize urban green space equity while considering four critical factors: individual mobility dynamics, neighborhood effects polarization, gender disparities, and vegetation chromatic characteristics.

1.4 Research Dimensions of UGS Equity

Social equity is a complex and multidimensional concept encompassing three core dimensions—procedural justice, distributive justice, and interactional justice—which are equally applicable to the equity of urban green spaces. Procedural justice emphasizes the fairness and transparency of decision-making and implementation processes; distributive justice focuses on the equitable allocation of resources, wealth, and opportunities; interactional justice prioritizes equality and respect in interpersonal relations, requiring the elimination of discrimination, bias, and unjust treatment in social interactions. These three dimensions are interconnected and mutually reinforcing, collectively forming the comprehensive framework of social equity.

1. Procedural Justice
Procedural justice denotes the rationality of decision-making processes in UGS allocation (spatially or quantitatively), emphasizing whether planning decisions incorporate diverse social needs and enable public participation before system planning, design, and implementation. UGS procedural justice emphasizes individualized treatment expectations in resource allocation—whether actual or anticipated—which materializes through three constitutive mechanisms: culturally responsive recognition during planning phases, structured involvement of social groups in governance (whether voluntary or institutionally mandated), and formalized institutional procedures enabling democratic contestation, negotiation, and ratification of green space policies. Achieving procedural green justice requires institutional designs that provide participatory platforms for marginalized communities. Cross-national variations in political systems necessitate context-specific institutional innovations to ensure equitable discursive representation.

2. Distributional Justice
Distributional justice refers to the fairness of spatial outcomes in UGS planning, emphasizing that UGS layouts must satisfy diverse social groups and guarantee residents' rights to enjoy corresponding green amenities. This represents the primary focus in most green equity literature. An idealized equitable distribution

would ensure universal availability of UGS with minimum usage rights for all individuals, yet such conditions remain largely unrealized in practice. Researchers typically examine concrete manifestations through two approaches: analyzing UGS as public resources allocated across statistically defined populations, or assessing spatial accessibility differentials for various social groups. Following identification of distribution patterns and existing deficiencies, urban planning interventions—such as optimizing street greening levels and controlling green space scale hierarchies—can improve spatial equity outcomes.

3. Interactional Justice
Interactional justice necessitates undifferentiated rights for all residents in accessing and utilizing UGS, mandating equitable treatment across social groups during UGS usage while establishing formal channels for feedback and rights protection regarding distributional injustices. This dimension manifests concretely through: (1) class-based disparities in accessing green services; (2) sensory experiences engendered by cultural symbols, landmarks, and spatial design languages within UGS environments; (3) interpersonal interaction quality or perceived environmental comfort during public green space usage; and (4) the efficacy of public oversight mechanisms regulating UGS service quality post-implementation. Consequently, achieving interactional justice requires institutionalizing standardized protocols for post-construction supervision, management, and maintenance alongside mechanisms ensuring their consistent implementation and iterative optimization.

Procedural, distributional, and interactional justice—representing distinct phases of UGS resource allocation—exhibit differentiated characteristics; their realization and enhancement necessitate embedding targeted institutional logics within green space planning frameworks, addressing unique governance challenges at each resource configuration stage.

1.5 Policy Depth of Urban Green Space Equity

1. Statutory Frameworks
The statutory framework serves as the legal foundation for planning formulation and implementation systems. Current regulations pertaining to UGS comprise two categories: advisory provisions and quantitative standards (Table 1.1). Advisory provisions establish qualitative requirements for UGS planning objectives without specifying measurable benchmarks, whereas quantitative standards involve explicit metrics subdivided into binding and indicative types. Binding standards—mandatory commitments by governments to guarantee public access to quality green environments—contrast with indicative standards, which express non-compulsory aspirational targets for future UGS development.

Within quantitative standards, binding metrics predominantly reflect aggregate or proportional indicators (e.g., per capita green space area, greening rate), often prioritizing spatial uniformity over equitable service distribution across diverse

Table 1.1 Green inequity in China's spatial planning regulatory framework

Law & regulation type	Name of law/regulation	Enactment (revision) year	Green space-related provisions	Provision type	Green equity issues
Law	Urban and Rural Planning Law of the People's Republic of China	2007 (2019)	Article 4: Improve the ecological environment, promote resource/energy conservation and comprehensive utilization, protect cultivated land and other natural/historical/cultural resources	Advisory	1. Lacks elaboration on defining "improved/protected" ecological environments or realized ecological value 2. Insufficient explicit mention of social equity aspects in green space planning
Administrative Regulation/Rule/Normative Document	Measures for the Formulation and Approval of Provincial Urban System Plans	2010	Article 4: Formulating provincial urban system plans shall... protect natural/cultural resources and the ecological environment	Advisory	
	Urban Planning Formulation Measures	2005	Article 4: Formulating urban plans shall... uphold resource conservation/intensive utilization, protect the ecological environment	Advisory	
			Article 14: During master plan formulation, major special topics involving resource/environmental protection... shall be researched by experts organized by municipal governments	Advisory	Lacks stipulations regarding public participation in planning formulation processes
	Urban Design Administration Measures	2017	Article 11: Urban design for key streets/blocks shall... expand pedestrian activity and greening space	Advisory	Definition/evaluation of area importance may inadequately consider social equity in practice
	13th Five-Year Plan for Ecological Environmental Protection	2016	Section 6: Expand ecological product supply... By 2020, urban per capita park green space ≥14.6 m^2, urban built-up area greening rate ≥38.9%	Target (Projected)	1. Projected targets lack requirements for green space quality enhancement 2. In practice, projected targets often function as de facto binding targets, leading to "quantity-over-quality" in green space planning/construction
Local Regulation/Ordinance	Beijing Urban and Rural Planning Regulations	2009 (2019)	Article 89: Legal authorities/organizations may file public interest litigation against illegal construction damaging ecology/resource protection to claim compensation	Advisory	Inadequate mechanisms may lead to unclear accountability

Technical Standard/ Code	Standard for Classification of Urban Land Use and Planning Construction Land (GB 50137-2011)	2011	4.3: Per capita urban land-use standards... Per capita green space ≥10.0 m²/capita (incl. park green space ≥8.0 m²/capita)	Binding Indicator	When stipulating metrics related to green space planning, solely adopting indicators such as per capita green space area and greening rate reflects only "equality" rather than "equity"
			4.4: Urban construction land structure... Green and square space proportion: 10.0%~15.0%	Binding Indicator	
	Standard for Urban Residential District Planning (GB 50180-2018)	2018	4.0.4: New residential districts... Public green space standards... 15-min living circle: Per capita public green space ≥2.0 m²/capita, District park min. size 5.0 ha/min. width 80 m	Binding Indicator	
			A.0.1.3: Adjacent to expressway/highway? Residential land boundary extends to road red line/buffer green space. Expressway/highway & buffer green space excluded from residential land	Advisory	Inflexible land boundary definitions may hinder integrated coordination of green space metrics across types
	Urban Green Space Planning Standard (GB/T 51346-2019)	2019	4.2.3: Per capita regional green space within urban development boundary ≥20 m²/capita	Binding Indicator	Lack of quantitative metrics reflecting social disparity dynamics
			4.2.8: Per capita scenic recreation green space ≥20 m²/capita (≥10 m²/capita in urban development boundary)	Binding Indicator	
			4.3.3: Park green spaces shall be distributed by service radius hierarchy	Advisory	Regulating green space allocation solely based on factors like service radius and scale overlooks actual needs of social groups, accessibility, and economic constraints
			5.1.2: Newly planned comprehensive parks >10 ha	Binding Indicator	
			5.1.6: Residential district parks >1 ha shall include: children's play, leisure, sports/fitness, cultural education, public services, commercial services, park administration facilities	Advisory	

populations. This approach fails to address differentiated accessibility needs among socio-demographic groups. Indicative standards, meanwhile, typically regulate quantitative parameters (e.g., total UGS area) while neglecting quality controls, incentivizing numerical target fulfillment at the expense of functional or experiential quality—a manifestation of procedural inequity in public resource allocation. Advisory provisions frequently mandate ecological conservation and resource efficiency yet lack operational specificity; their ambiguous implementation mechanisms and undefined accountability structures create governance gaps, including unclear liability attribution for UGS rights infringements and insufficient enforcement of planning outcome evaluations.

2. Administrative Frameworks

The planning administrative system comprises vertical hierarchies (multi-tiered departments overseeing spatial planning at corresponding scales) and horizontal networks (inter-departmental coordination at equivalent governmental levels). Vertical frameworks entail supervisory guidance from superior to subordinate authorities, while horizontal systems require collaborative governance among peer agencies. Critically, vertical dysfunctions manifest as insufficient policy transmission across tiers, whereas horizontal fragmentation reflects ambiguous accountability across departments.

Within vertical governance, intergovernmental transmission of planning elements—including quantitative metrics, spatial structures, regulatory inventories, locational parameters, zoning classifications, and boundary demarcations—often engenders procedural green injustice. This occurs through two primary mechanisms: (1) Deficient metric transmission: Quantitative standards for UGS allocation (e.g., per capita green space) lack binding mechanisms in built-environment planning frameworks. Crucially, internal green space metrics within construction land parcels are excluded from compulsory transmission protocols, depriving UGS planning of vertical regulatory constraints. (2) Fragmented boundary implementation: The demarcation of "Three Zones and Three Lines" within territorial spatial planning requires top-down refinement through multiple tiers: structural boundary delineation is executed at national and provincial planning levels, preliminary boundary definition occurs at municipal/county planning stages, and precise spatial implementation is ultimately achieved through township-level territorial spatial plans or detailed development regulations.

Horizontally, overlapping jurisdictional voids persist despite the 2018 establishment of the Ministry of Natural Resources (MNR) consolidating spatial planning mandates. While UGS planning now resides within MNR-administered territorial spatial plans, operational responsibilities—including municipal facilities management, landscape engineering, and environmental maintenance—remain under the Ministry of Housing and Urban-Rural Development (MOHURD). This institutional bifurcation creates accountability ambiguities in post-implementation UGS governance, fostering inter-departmental responsibility displacement that systematically hindering UGS implementation efficiency.

3. Plan Formulation Systems

The spatial planning hierarchy operates through tiered formulation systems, where each administrative level develops master plans, detailed regulations, and specialized plans. The formulation process—encompassing drafting, public consultation, and revision phases—systematically engenders procedural green injustice due to structural deficiencies in civic engagement. This manifests through four critical failures: (1) Marginalization of participatory scoping: Planning entities routinely neglect to identify heterogeneous community demands during preliminary field investigations, resulting in unrecorded disparities in green space allocation claims. (2) Constricted engagement channels: Public hearings and consultations function as notification-based participation mechanisms, wherein restricted eligibility criteria and procedural complexities inherently limit stakeholder inclusion to passive receipt of predetermined decisions. (3) Accessibility barriers in plan dissemination: Published planning documents deploy hyper-technical lexicons without supplementary visualizations or simplified interpretations, effectively nullifying public comprehension and undermining statutory rights to informed participation. (4) Accountability voids in feedback loops: The absence of mandatory disclosure protocols for recording, processing, and incorporating public input perpetuates civic disillusionment, as citizens remain uninformed regarding the adoption status of their proposals—a condition exacerbating perceptions of institutional disregard and procedural inequity.

4. Implementation Systems

The implementation supervision framework ensures effective execution of formulated green space plans through three institutional components: planning implementation organization, regulatory management, and public oversight mechanisms. Systemic distributive inequities emerge across these dimensions due to institutionalized fiscal biases and accountability deficits.

Implementation Organization manifests procedural green injustice as governmental agencies prioritize high-visibility megaprojects through backward-planning approaches for political performance metrics, systematically relegating green infrastructure investments to secondary status. Concurrently, fiscal capacity variances among municipalities—driven by divergent socioeconomic conditions—create spatial disparities in public green space funding, directly constraining equitable access to quality urban ecosystems for marginalized communities.

Regulatory Management reveals inherent distortions as the land-dependent fiscal model incentivizes local governments to commercialize public land through transfer mechanisms rather than welfare-oriented allocation. These trigger cascading effects: systematic underfunding of green space quotas, encroachment on existing ecological corridors, and intensified socioeconomic inequities through gentrification-driven farmland conversion, collectively undermining distributive justice in urban green resource allocation.

Public Oversight Mechanisms remain structurally deficient, characterized by opaque monitoring protocols and restricted civic participation channels. The absence of statutory grievance redress procedures perpetuates interactional injustice—when green space projects face indefinite delays or cancellations, affected communities lack legal recourse to challenge rights violations, resulting in prolonged exposure to institutional neglect and eroding public trust in planning governance.

1.6 Characterization Metrics for Urban Green Space Equity

When measuring urban green space equity, it typically requires the assessment of green space exposure characteristics alongside socioeconomic and demographic attributes, as well as evaluating the level of equity manifested by these green exposure features.

1. Characterization Metrics of Urban Green Space

Urban green space characterization involves assessing three core dimensions: quantitative attributes measure green space area and patch scale; qualitative attributes evaluate aesthetic and ecological values such as landscape scenic integrity and biodiversity indices; functional attributes analyze recreational services including spatial facility configurations, pedestrian accessibility, and adjacent land-use patterns. Substantial empirical evidence supports correlations between these metrics and spatial equity outcomes, establishing a robust methodological foundation for this research.

Regarding data sources, green space metrics derive from both quantitative and qualitative approaches. Quantitative analysis utilizes remote sensing imagery and open-source geospatial data to calculate landscape ecology indices (e.g., surrounding green space area, vegetation coverage rates, and landscape shape index), offering objective measurement advantages while constrained by limited small-scale feature resolution. Qualitative methods employ self-reported surveys and trained-observer audits to document usage patterns including visit frequency, duration, and activity types, yielding comprehensive behavioral insights at higher resource cost. Emerging hybrid strategies integrate geo-big data with targeted field sampling to resolve the inherent tension between broad geographical coverage and granular spatial precision.

Green space exposure modeling historically depended on accessibility simulations to approximate visitation behaviors, partially differentiating objective green infrastructure from actual exposure contexts. Despite iterative refinements, these models still struggle to reflect true exposure dynamics. Inherent methodological challenges include geographical contextual uncertainty, residential self-selection bias, mobility-induced distortions, and time-lag effects—all contributing to biased exposure estimations. Traditional studies emphasize neighborhoods as static environments significantly influencing resident health. However, workplaces serve as critical anchors where prolonged exposure occurs, potentially compensating for residential green space deficiencies. Empirical research consistently reveals disparities between static geographical environment (residential/work locations) and dynamic exposure environment (mobility corridors), while cumulative contact effects remain understudied. Daily mobility variations and urban population density fluxes critically impact assessment accuracy, necessitating prioritized development of real-time dynamic exposure monitoring techniques for urban green space research.

2. Socioeconomic and Demographic Metrics

As urban green spaces evolve to embody both social and commodity attributes within urbanization processes, discussions on their equity intrinsically intersect with the socioeconomic characteristics of study units. These attributes concretely manifest through human-centric metrics such as economic capacity, social status, age, gender, and ethnicity, as well as broader development indicators specific to cities, nations, or regions. Empirical investigations across distinct socioeconomic groups consistently identify significant green inequality patterns.

1. Socioeconomic capacity. Socioeconomic capacity manifests primarily through individual/household income and assets, constituting the most extensively debated attribute in green equity discourse. Extensive empirical studies confirm disparities in urban green space accessibility across communities of varying socioeconomic status (SES). Compared to high-income cohorts, low-income neighborhoods exhibit significantly reduced accessibility to urban green spaces, with proximate green infrastructure typically characterized by smaller scale and higher crowding density. Concurrently, funding allocations from national and non-profit organizations disproportionately favor middle-to-high-income communities. Furthermore, newly developed green spaces demonstrate locational bias toward affluent neighborhoods, thereby attracting higher-income migrants and exacerbating spatial inequities in urban green resource distribution.
2. Age. Age intrinsically correlates with mobility capacity, particularly impacting children and elderly populations who encounter movement constraints when accessing urban green spaces, while exhibiting distinct demand preferences. Elderly individuals prioritize tranquil, hygienic environments with ample shaded seating areas, whereas younger demographics favor spaces accommodating physical activities. Failure to accommodate these age-specific demand differentials in urban green space design directly precipitates distributional inequity.
3. Gender. Gender-based disparities in urban green space equity constitute a relatively peripheral yet critical discourse. Physiological and psychological distinctions between sexes, coupled with differential access to opportunities versus exposure to constraints in social life, engender divergent patterns in green space utilization and perceptual experiences. In many nations, the systemic subordination of women within social hierarchies frequently precludes their participation in green space decision-making processes, thereby institutionalizing inequitable outcomes.
4. Race, Religion, and Social Identity. Racial and religious factors, particularly salient in Western scholarship, reveal profound spatial injustices. Communities predominantly inhabited by racial minorities demonstrate markedly inferior green space accessibility, with proximate green infrastructure characterized by

heightened crowding intensity. Crucially, the planning and design of urban green spaces rarely integrate the specific needs of ethnic minority groups. Beyond physical exclusion, minority populations may encounter racial profiling within green spaces, where these areas paradoxically transform into sites of criminal activity, social conflict, and racial segregation. The complexity of socio-political histories further compounds green injustice through identity-based stratification. Disparities rooted in social identities—manifested across interstate relations, citizen-state dynamics, and central-local governance tensions—generate heterogeneous claims to green space entitlements. Immigrant communities and populations bearing colonial legacies frequently experience compromised access to green services, reflecting systemic privileging of dominant identity groups within spatial resource allocation frameworks.

5. Socio-demographic Development Level. Socio-demographic development metrics—including the Human Development Index (HDI), social demographic indices, SES, population growth rates, and illiteracy rates—serve as critical indicators for quantifying developmental disparities across cities, nations, and regions. Divergent climatic conditions, historical trajectories, and cultural paradigms fundamentally shape heterogeneous patterns of urban green space equity. As urban planning capacity, demographic demands, and wealth accumulation intensify with urban development, four interconnected vectors—urban spatial governance, green infrastructure conservation, wealth inequality, and market interventions—collectively reconfigure systemic equity in green resource allocation. Empirical studies consistently document superior green equity outcomes in Global North cities relative to their Global South counterparts. Crucially, equitable green space exposure may manifest in both resource-deprived and resource-abundant urban contexts, though their underlying mechanisms diverge significantly: in disadvantaged cities, equity often stems from policy-driven redistribution, whereas in affluent settings it frequently reflects market-induced homogenization of access.

Socio-demographic economic characteristics can be categorized into fundamental attributes and behavioral dynamics. While questionnaires remain a prevalent measurement tool, they are persistently constrained by subjective misreporting biases. To enhance precision, integrated methodologies combining surveys, behavioral trajectory tracking, and travel diaries have been developed—yet large-scale deployment faces prohibitive costs due to geographical coverage limitations. Consequently, interpretive techniques leveraging crowdsourced big data analytics are emerging as scalable alternatives. Mobile-signaling data converging with travel software platforms (e.g., Liangbulu app for trail recording, Foooooot app for GPS-based navigation communities) collectively enable machine learning-driven identification of residential, occupational, and recreational zones through their recorded user profiles and mobility trajectories. Concurrently, movement velocities and origin-destination matrices reveal comprehensive mobility signatures, establishing a robust sample information database for behavioral analytics. Socioeconomic statistical data demonstrates heightened accessibility across multi-tier administrative units. Population censuses and local statistical yearbooks contain granular metrics encompassing GDP stratification, industrial composition, demographic distributions, and educational gradients, establishing foundational references for China-specific urban research. Satellite-acquired nighttime light indices correlate significantly with urban economic development, providing globally continuous longitudinal datasets for planning retrospection. Crucially, geospatial big data platforms—including WorldPop, LandScan, GPW (Gridded Population of the World), China's Kilometer-grid Population Distribution Database, and GRUMP (Global Rural-Urban Mapping Project)—are extensively leveraged to quantify urban demographic-economic development patterns through high-resolution spatial disaggregation and dynamic modeling algorithms.

3. Analytical Framework for Green Equity Assessment
In green equity research, statistical and econometric methodologies constitute essential tools for quantifying disparities in green resource allocation across socioeconomic groups. These approaches fundamentally seek to detect differentials in green space exposure characteristics among populations of varying SES. Scholars have developed multiple analytical frameworks for such assessments, notably:

1. Gini Index and Lorenz Curve constitute classical metrics for assessing spatial inequality. By computing the distributional differentials of green spaces (e.g., parks, urban greenery) across distinct geographical units or demographic groups, the Gini index quantifies the magnitude of inequitable green resource allocation. The Lorenz Curve, as a graphical representation, visually delineates cumulative distribution patterns, providing an intuitive reflection of distributive fairness in green resource accessibility. Furthermore, supplementary inequality indices—including the Theil Index, Hoffmann Index (Hoover Index), and Atkinson Index—are deployed to measure multidimensional equity dimensions.
2. The Pearson correlation coefficient serves as a fundamental metric for quantifying associations between green resource distribution patterns and socioeconomic determinants. For instance, investigations into urban park equity typically employ calculated correlations between park accessibility indices and socioeconomic metrics—such as income stratification gradients and elderly population ratios—to systematically unmask underlying allocation inequities.
3. Spatial autocorrelation analysis methodologies, exemplified by Moran's I, quantify spatial agglomeration characteristics within green resource distribution systems. Applied to urban park equity investigations, this index operationalizes the detection of spatially clustered allocation patterns, thereby pinpointing geographically explicit deprivation zones where statistical evidence of resource imbalance is confirmed through clustering signatures.
4. Regression modeling statistically verifies associations between dependent variables (green exposure metrics) and independent variables (socioeconomic attributes), thereby quantifying their joint impacts on green equity. This approach enables confounding factor control while detecting mediation or interaction effects among multivariate drivers, ultimately unmasking complex causal mechanisms underlying environmental inequalities.

5. Machine learning methodologies excel at processing complex data structures and high-dimensional features, particularly in capturing non-linear dependencies between green exposure metrics and socioeconomic characteristics. A representative case manifests in urbanization trajectories: initial built-up area expansion typically constricts green space development, whereas subsequent institutional refinements and capital investments gradually elevate green infrastructure quality. This dynamic often generates an inverted U-shaped trajectory in green equity—progressing from escalating disparity toward progressive equalization—a phenomenon effectively investigable through Bayesian quantile regression frameworks.

In summary, while differential methodological tradeoffs exist between individual attribute quantification and socioeconomic metrics acquisition, the progressive integration of big data and artificial intelligence technologies is driving systematic resolution of conventional limitations. Advancing computational capacities and novel technological paradigms will therefore propel behavioral analytics and socioeconomic statistics toward progressively refined precision and operational efficiency, ultimately furnishing urban governance frameworks and policy formulation with empirically grounded foundations that demonstrably accelerate sustainable societal development.

1.7 The Multi-Scale Analytical Framework of This Book

On the basis of the foregoing literature review, this book proposes a multi-scalar analytical framework for urban green space equity oriented toward the "nation–city–individual" continuum. Environmental justice's tripartite structure—procedural, distributive, and interactional justice—serves as the theoretical spine, embedded within China's policy and governance context.

Conceptually, the framework classifies study variables into four families. First are supply-side factors: green space availability and visibility, accessibility and service capacity, and spatial–morphological characteristics. Second are demand-side factors: individual attributes such as income, age, gender, and ethnicity/social identity, alongside city- and region-level socioeconomic indicators and built-environment variables. Third are processual factors, encompassing institutional and governance processes as well as residents' daily mobility and spatiotemporal behavior. Fourth are outcome factors—multidimensional representations and spatial patterns of green space equity—including overall inequality captured by the Gini index and its extensions, deprivation clusters identified via local spatial autocorrelation and differentiated patterns across social groups and administrative units.

Methodologically, the framework emphasizes a "scale matching and indicator consistency" pathway to balance international comparability with contextual fit for China. At the national scale, built-up areas constitute the basic units of analysis. Remote sensing is integrated with navigation/map big data to construct a three-part "green exposure" comprising resource endowments, residential segregation, and public green space accessibility. Inequality related to income status is profiled using a generalized Gini index based on price-ranked cumulative green exposure, while spatial autocorrelation and multivariate regressions identify drivers and regional heterogeneity. At the city scale, patch-based greens and linear street greenery are unified within a Comprehensive Floating Catchment Area (CFCA) model that explicitly incorporates multiple travel modes, distance-decay, supply–demand coupling, and scale-specific service radii; parallel "quantity-based" and "quality-based" accessibility measures are used to test the bias that arises when quantity is treated as a proxy for quality. At the individual scale, GPS traces, travel diaries, and questionnaires link static residential/workplace exposure with dynamic travel-based exposure, allowing assessment of the conditions and mechanisms under which compensatory effects or polarization occur—thereby correcting mismeasurement that arises when exposure is inferred from residence alone. To characterize nonlinearities and interactions, the framework employs interpretable machine learning to quantify thresholds and coupling effects of spatial-morphology indicators on equity outcomes, with classical econometric models providing robustness checks.

In data and measurement, the framework advocates multi-source fusion and vertical alignment across scales: remote sensing imagery and nighttime lights to reveal ecological and developmental conditions; open maps and street-view imagery, processed via deep learning, to extract integrated patch–linear–point green space features; navigation APIs to supply real-world networks and travel times; population censuses and statistical yearbooks to profile urban socioeconomic structures; and community/estate web-scraped data to supplement residential attributes and price gradients. All indicators are normalized to ensure comparability and are aligned with internationally recognized evaluation conventions connected to SDG 11, enabling China–global benchmarking and context-sensitive interpretation in subsequent chapters.

For explanation and translation to practice, the framework establishes a policy loop of "diagnosis–prioritization–intervention–evaluation." Inequity types and hotspots are first identified using the Gini index and spatial clustering; SHAP-based importance and threshold rules then generate structured intervention guidance; heuristic methods sequence and optimize policies such as opening residential compounds. In parallel, procedural justice is operationalized through institutionalized public participation, clarified inter-departmental accountability, and recalibrated technical standards to mend transmission gaps from regulation to implementation. Beyond quantitative expansion, the framework highlights active-travel-friendly design and enhancement of linear green infrastructure as primary pathways for quality improvement, seeking co-benefits across ecology, health, and social justice.

The book is organized as follows: Chap. 2 operationalizes the triadic green exposure and inequality metrics at the national level and explains how administrative hierarchy and regional differentiation systematically shape equity; Chap. 3 conducts parallel quantity- and quality-based accessibility assessments and policy simulations at the city level, testing the actual distributive effects of integrated patch–corridor–street systems and "open community" strategies; Chap. 4 details biases and mechanisms in static–dynamic exposure at the individual level, arguing for the inclusion of workplaces and travel paths in equity assessments; Chap. 5 synthesizes multi-scalar evidence, benchmarks against international scholarship, and proposes China-specific institutional reforms and a future research agenda. Through this design, the framework addresses common global concerns in green space equity while offering measurement tools and governance pathways that are operational within China's distinctive institutional and developmental context.

References

Blum ED (2008) Love Canal revisited: Race, class, and gender in environmental activism. University Press of Kansas, Lawrence

Boone CG, Buckley GL, Grove JM, Sister C (2009) Parks and people: an environmental justice inquiry in Baltimore, Maryland. Ann Assoc Am Geogr 99(4):767–787. https://doi.org/10.1080/00045600903102949

Bryant B (1995) Environmental justice: issues, policies, and solutions. Island Press, Washington, DC

Burgess J, Harrison CM, Limb M (1988) People, parks and the urban green: a study of popular meanings and values for open spaces in the city. Urban Stud 25(6):455–473

Byrne J (2012) When green is white: the cultural politics of race, nature and social exclusion in a Los Angeles urban national park. Geoforum 43(3):595–611. https://doi.org/10.1016/j.geoforum.2011.10.002

Chen Y, Yue W (2020) Which communities have better accessibility to green space? An investigation into environmental inequality using big data. Landsc Urban Plan 204:103919. https://doi.org/10.1016/j.landurbplan.2020.103919

Chen Y, Xu Z, Byrne J, Xu T, Wang S, Wu J (2021) Can smaller parks limit green gentrification? Insights from Hangzhou, China. Urban For Urban Green 59:127009. https://doi.org/10.1016/j.ufug.2021.127009

Chen B, Tu Y, Wu S, Song Y, Jin Y, Webster C, Xu B, Gong P (2022a) Beyond green environments: Multi-scale difference in human exposure to greenspace in China. Environ Int. https://doi.org/10.1016/j.envint.2022.107348

Chen B, Wu S, Song Y, Webster C, Xu B, Gong P (2022b) Contrasting inequality in human exposure to greenspace between cities of Global North and Global South. Nat Commun 13(1):Article 1. https://doi.org/10.1038/s41467-022-32258-4

Chick G (2005) Are prejudice and discrimination learned? A comment on Stodolska. Leis Sci 27(1):41–43. https://doi.org/10.1080/01490400590886006

Diener A, Mudu P (2021) How can vegetation protect us from air pollution? A critical review on green spaces' mitigation abilities for air-borne particles from a public health perspective—with implications for urban planning. Sci Total Environ 796:148605. https://doi.org/10.1016/j.scitotenv.2021.148605

Dony CC, Delmelle EM, Delmelle EC (2015) Re-conceptualizing accessibility to parks in multimodal cities: A Variable-width Floating Catchment Area (VFCA) method. Landsc Urban Plan 143:90–99. https://doi.org/10.1016/j.landurbplan.2015.06.011

Edwards PK (1981) Race, residence, and leisure style: Some policy implications. Leis Sci 4(2):95–112. https://doi.org/10.1080/01490408109512954

Emmett AR (2015) Roman law, private property and the public domain: lessons for copyright policy. In: Fitzgerald B, Gilchrist J (eds) Copyright perspectives: past, present and prospect. Springer, Berlin, pp 17–27. https://doi.org/10.1007/978-3-319-15913-3_3

Estes S (2000) "I am a man!": Race, masculinity, and the 1968 Memphis Sanitation Strike. Labor Hist 41(2):153–170. https://doi.org/10.1080/00236560050009914

Fan P, Xu L, Yue W, Chen J (2017) Accessibility of public urban green space in an urban periphery: The case of Shanghai. Landsc Urban Plan 165:177–192. https://doi.org/10.1016/j.landurbplan.2016.11.007

Fernández IC, Wu J (2016) Assessing environmental inequalities in the city of Santiago (Chile) with a hierarchical multiscale approach. Appl Geogr 74:160–169. https://doi.org/10.1016/j.apgeog.2016.07.012

Gunawardena KR, Wells MJ, Kershaw T (2017) Utilising green and bluespace to mitigate urban heat island intensity. Sci Total Environ 584–585:1040–1055. https://doi.org/10.1016/j.scitotenv.2017.01.158

Johnson CY, Bowker JM, English DBK, Worthen D (1998) Wildland recreation in the rural south: an examination of marginality and ethnicity theory. J Leis Res 30(1):101–120. https://doi.org/10.1080/00222216.1998.11949821

Kim Y, Corley EA, Won Y, Kim J (2023) Green space access and visitation disparities in the phoenix metropolitan area. Landsc Urban Plan 237:104805. https://doi.org/10.1016/j.landurbplan.2023.104805

Kneebone E, Garr E (2010) The suburbanization of poverty. Brookings Institute, Washington, DC

Knobel P, Maneja R, Bartoll X, Alonso L, Bauwelinck M, Valentin A, Zijlema W, Borrell C, Nieuwenhuijsen M, Dadvand P (2021) Quality of urban green spaces influences residents' use of these spaces, physical activity, and overweight/obesity. Environ Pollut 271:116393. https://doi.org/10.1016/j.envpol.2020.116393

Kronenberg J, Haase A, Łaszkiewicz E, Antal A, Baravikova A, Biernacka M, Dushkova D, Filčak R, Haase D, Ignatieva M, Khmara Y, Niţă MR, Onose DA (2020) Environmental justice in the context of urban green space availability, accessibility, and attractiveness in postsocialist cities. Cities 106:102862. https://doi.org/10.1016/j.cities.2020.102862

Krymkowski DH, Manning RE, Valliere WA (2014) Race, ethnicity, and visitation to national parks in the United States: Tests of the marginality, discrimination, and subculture hypotheses with national-level survey data. J Outdoor Recreat Tour 7–8:35–43. https://doi.org/10.1016/j.jort.2014.09.008

LaCour-Little M (2001) Gated communities and property values. Wells Fargo Home Mortgage and Department of Real Estate and Urban Land Economics - University of Wisconsin, Madison, 36 p

Laituri M, Kirby A (1994) Finding fairness in America's cities? The search for environmental equity in everyday life. J Soc Issues 50(3):121–139. https://doi.org/10.1111/j.1540-4560.1994.tb02423.x

Lee RE, Booth KM, Reese-Smith JY, Regan G, Howard HH (2005) The Physical Activity Resource Assessment (PARA) instrument: Evaluating features, amenities and incivilities of physical activity resources in urban neighborhoods. Int J Behav Nutr Phys Act 2(1):13. https://doi.org/10.1186/1479-5868-2-13

Lu Y, Chen R, Chen B, Wu J (2024) Inclusive green environment for all? An investigation of spatial access equity of urban green space and associated socioeconomic drivers in China. Landsc Urban Plan 241:104926. https://doi.org/10.1016/j.landurbplan.2023.104926

Mitchell G, Norman P, Mullin K (2015) Who benefits from environmental policy? An environmental justice analysis of air quality change in Britain, 2001–2011. Environ Res Lett 10(10):105009. https://doi.org/10.1088/1748-9326/10/10/105009

Mohai P, Bryant B (1992) Environmental racism: reviewing the evidence. In: Race and the incidence of environmental hazards. Routledge, Boulder

Novotny P (2000) Where we live, work, and play: The environmental justice movement and the struggle for a new environmentalism. Greenwood, London

Opotow S, Clayton S (1994) Green justice: conceptions of fairness and the natural world. J Soc Issues 50(3):1–11. https://doi.org/10.1111/j.1540-4560.1994.tb02416.x

Rigolon A (2016) A complex landscape of inequity in access to urban parks: A literature review. Landsc Urban Plan 153:160–169. https://doi.org/10.1016/j.landurbplan.2016.05.017

Schlosberg D (2013) Theorising environmental justice: The expanding sphere of a discourse. Environ Polit 22(1):37–55. https://doi.org/10.1080/09644016.2013.755387

Schweizer E (1999) Environmental justice: An interview with Robert Bullard. Earth First! J 19(7):8

Sister C, Wolch J, Wilson J (2010) Got green? Addressing environmental justice in park provision. GeoJournal 75(3):229–248. https://doi.org/10.1007/s10708-009-9303-8

Smith SLJ (1980) Intervening opportunities and travel to urban recreation centers. J Leis Res 12(4):296–308. https://doi.org/10.1080/00222216.1980.11969456

Song Y, Chen B, Ho HC, Kwan M-P, Liu D, Wang F, Wang J, Cai J, Li X, Xu Y, He Q, Wang H, Xu Q, Song Y (2021) Observed inequality in urban greenspace exposure in China. Environ Int 156:106778. https://doi.org/10.1016/j.envint.2021.106778

Stodolska M (2005) Implications of the conditioned attitude model of individual discriminatory behavior for discrimination in leisure settings. Leis Sci 27(1):59–74. https://doi.org/10.1080/01490400590886051

Stodolska M, Jackson EL (1998) Discrimination in leisure and work experienced by a white ethnic minority group. J Leis Res 30(1):23–46. https://doi.org/10.1080/00222216.1998.11949817

Stoep GAV (1993) Proceedings of the 1993 Northeastern recreation research symposium: April 18–20, 1993, State Parks Management and Research Institute, Saratoga Spa State Park, Saratoga Springs, New York. U.S. Department of Agriculture, Forest Service, Northeastern Forest Experiment Station

Talen E (2003) Neighborhoods as service providers: a methodology for evaluating pedestrian access. Environ Plann B Plann Des 30(2):181–200. https://doi.org/10.1068/b12977

Tian M, Yuan L, Guo R, Wu Y, Liu X (2021) Sustainable development: Investigating the correlations between park equality and mortality by multilevel model in Shenzhen, China. Sustain Cities Soc 75:103385. https://doi.org/10.1016/j.scs.2021.103385

Vaughan KB, Kaczynski AT, Wilhelm Stanis SA, Besenyi GM, Bergstrom R, Heinrich KM (2013) Exploring the distribution of park availability, features, and quality across Kansas City, Missouri by income and race/ethnicity: an environmental justice investigation. Ann Behav Med 45(suppl_1):S28–S38. https://doi.org/10.1007/s12160-012-9425-y

Wang R, Cao M, Yao Y, Wu W (2022) The inequalities of different dimensions of visible street urban green space provision: A machine learning approach. Land Use Policy 123:106410. https://doi.org/10.1016/j.landusepol.2022.106410

Wang J, Kwan M-P, Xiu G, Peng X, Liu Y (2024) Investigating the neighborhood effect averaging problem (NEAP) in greenspace exposure: A study in Beijing. Landsc Urban Plan 243:104970. https://doi.org/10.1016/j.landurbplan.2023.104970

Washburne RF (1978) Black under-participation in wildland recreation: Alternative explanations. Leis Sci 1(2):175–189. https://doi.org/10.1080/01490407809512880

Wolch J, Wilson JP, Fehrenbach J (2005) Parks and park funding in Los Angeles: an equity-mapping analysis. Urban Geogr 26(1):4–35. https://doi.org/10.2747/0272-3638.26.1.4

Wolch JR, Byrne J, Newell JP (2014) Urban green space, public health, and environmental justice: The challenge of making cities 'just green enough'. Landsc Urban Plan 125:234–244. https://doi.org/10.1016/j.landurbplan.2014.01.017

Wu L, Kim SK (2021) Exploring the equality of accessing urban green spaces: A comparative study of 341 Chinese cities. Ecol Indic 121:107080. https://doi.org/10.1016/j.ecolind.2020.107080

Wu J, Feng Z, Peng Y, Liu Q, He Q (2019) Neglected green street landscapes: A re-evaluation method of green justice. Urban For Urban Green 41:344–353. https://doi.org/10.1016/j.ufug.2019.05.004

Wu J, Chen H, Wang H, He Q, Zhou K (2020a) Will the opening community policy improve the equity of green accessibility and in what ways?—Response based on a 2-step floating catchment area method and genetic algorithm. J Clean Prod 263:121454. https://doi.org/10.1016/j.jclepro.2020.121454

Wu J, Wang B, Ta N, Zhou K, Chai Y (2020b) Does street greenery always promote active travel? Evidence from Beijing. Urban For Urban Green 56:126886. https://doi.org/10.1016/j.ufug.2020.126886

Wu J, Xu Z, Jin Y, Chai Y, Newell J, Ta N (2022) Gender disparities in exposure to green space: An empirical study of suburban Beijing. Landsc Urban Plan 222:104381. https://doi.org/10.1016/j.landurbplan.2022.104381

Xu C, Haase D, Pribadi DO, Pauleit S (2018) Spatial variation of green space equity and its relation with urban dynamics: A case study in the region of Munich. Ecol Indic. https://doi.org/10.1016/J.ECOLIND.2018.05.024

Yan L, Jin X, Zhang J (2024) Equity in park green spaces: A bibliometric analysis and systematic literature review from 2014-2023. Front Environ Sci 12. https://doi.org/10.3389/fenvs.2024.1374973

Zhang L, Chen P, Hui F (2022) Refining the accessibility evaluation of urban green spaces with multiple sources of mobility data: A case study in Shenzhen, China. Urban For Urban Green 70:127550. https://doi.org/10.1016/j.ufug.2022.127550

Chapter 2
Research Perspective on National Scale of Urban Green Space and Equity

2.1 Cross-National Disparities in Urban Green Space Equity

Although varying degrees of green inequality have been documented in cities worldwide, recent studies have reported contradictory findings across geographical regions, developmental stages, and even within individual cities (Chen and Yue 2020; Lu et al. 2024; Xiao et al. 2017). Two primary factors may account for these discrepancies:

On one hand, as previously noted, these studies measure distinct dimensions of UGS. For instance, when assessing the same city, different scholars have employed vegetation coverage rate (%GA), green accessibility, and residential greening rate as metrics—yielding divergent conclusions ranging from severe green inequity, partial inequity, to no significant equity issues (Fan et al. 2017; Fung and Siu 2001; Li et al. 2015; Qiao et al. 2021). On the other hand, the studied cities operate at disparate socioeconomic stages. Cross-city studies indicate that green equity levels may be significantly influenced by the interconnected effects of socioeconomic indicators and environmental factors. Socioeconomic indicators (e.g., GDP) can substantially impact green equity by shaping governmental fiscal expenditures on green space construction, management, and maintenance—thereby affecting urban green space planning and the enforcement of conservation measures (Wu and Kim 2021). Cities at different socioeconomic stages exhibit variations in population size and industrial structure, which may generate reciprocal or antagonistic effects with urban green spaces, ultimately influencing their spatial distribution patterns.

J. Wu, *Urban Green Space Equity in China*, Urban Sustainability,
https://doi.org/10.1007/978-981-95-5771-4_2

2.1.1 National-Scale Dimensions of Green Equity

The urban green space metrics in green equity research can be categorized into three dimensions driving green inequality: residential segregation, resource endowments, and urban planning (Bolitzer and Netusil 2000; Panduro and Veie 2013). These three dimensions essentially correspond to distinct green space typologies—residential green spaces as private commodities, peri-residential green spaces as quasi-public goods, and public urban green spaces as common-pool resources. While the differential impacts of economic development stages on these green space types remain underexplored, their governance mechanisms and equity outcomes are inherently divergent.

Residential segregation often triggers environmental equity issues. Since the late twentieth century, China's economic transition has accelerated housing marketization and intensified residential segregation (Wang et al. 2011). Currently, the proliferation of gated communities allows residents of high-end housing to monopolize premium green spaces, exacerbating green inequity (Fleischer 2007). Cities at advanced economic development stages exhibit higher marketization levels, implying greater diversity in commodity supply and reduced government regulation over private goods provision (Bai et al. 2012; Chen et al. 2016). This suggests that in contexts of residential segregation, economically advanced cities may supply residential areas of more varied quality, potentially amplifying social inequality. However, when market competition is sufficiently intense, the abundance of diverse, high-quality commodities could enable residents with lower socioeconomic status to access communities with high greening rates, thereby achieving green equity.

Disparities in residential environmental resource endowments also influence residents' green access (Kronenberg et al. 2020; Wang et al. 2021). The phenomenon of green gentrification implies that socioeconomically advantaged groups often relocate to suburbs for superior green resources (Koprowska et al. 2020). However, studies indicate that while large green spaces may trigger gentrification, small-scale street greening in urban areas can mitigate this effect. Economic development typically enables greater potential funding for greening construction and maintenance, thereby elevating urban green resource endowments (Chen et al. 2021; Wang et al. 2022). Yet, this pattern is not globally universal—a U.S. study found urban green coverage declines with economic growth. China's urban development and public green space governance diverge significantly from Western models. Inter-city competition incentivizes Chinese local governments: those awarded titles like "National Forest City," "National Garden City," or "National Eco-City" gain substantial political rewards. Consequently, alongside pursuing economic growth, governments exhibit stronger motivation to construct or upgrade green spaces near residential areas—even in aging neighborhoods—enhancing equity in resource endowments.

Urban planning-induced green equity issues are particularly evident in Public Green Space (PGS) accessibility (Rigolon 2016). PGS planning in China is predominantly led by municipal governments, yet its spatial allocation is often influenced by capital and market forces under fiscal constraints, especially in developing contexts. China's unique land finance-driven urbanization model empowers local governments to capitalize urban land through residential and commercial land leasing, fueling infrastructure development. However, market-oriented planning and short-term economic gains have triggered the conversion of green spaces to other uses, amplified capital's dominance in greening decisions (Choumert and Salanié 2008), and systematically deprived socioeconomically disadvantaged groups of PGS accessibility. Existing literature highlights the negative correlation between economic development and PGS equity. Cities at advanced socioeconomic stages exhibit heavier reliance on land finance—a double-edged sword: On one hand, this extra-budgetary revenue funds urban infrastructure, stimulates labor and material markets, and propels local economic growth (Lichtenberg and Ding 2009; Yew 2012). On the other hand, such dependence is sustainable only under high-speed economic expansion; declining growth rates would jeopardize debt repayment for infrastructure investments (Wu and Kim 2021). Consequently, local governments prioritize economic agendas over equitable urban planning, frequently sacrificing preexisting green space configurations to accommodate growth-oriented industrial layouts or attract capital inflows.

Therefore, this study will examine built-up areas of cities nationwide as the research object. Through three dimensions—residential segregation, resource endowments, and urban planning—we conduct an in-depth analysis of the equality in green exposure accessibility among residents across socioeconomic strata in China. Furthermore, we investigate disparities in green equity across cities at varying development stages and among different green space dimensions. Specifically: residential segregation is quantified by community greening rate; resource endowments are measured using the NDVI; urban planning interventions are evaluated via accessibility metrics.

2.1.2 *Measurement Methods for Urban Green Spaces and Their Equity at the National Scale*

The boundary of the urban built environment in 2020 was extracted from the National Land-Use/Cover Database for China (NLUD-C), a dataset created by the Chinese Academy of Sciences that features a spatial resolution of 30 m. Derived from Landsat TM images and China-Brazil Earth Resources Satellite images, the NLUD-C encapsulates land-use coverage information dating back to 1990 (https://www.resdc.cn). Rigorous sampling procedures have validated the NLUD-C's classification accuracy, which exceeded 90%. Subsequently, we selected 263 cities boasting sufficiently comprehensive data for our study, encompassing 4 municipalities, 14 sub-provincial cities, 16 provincial capitals, and 229 prefecture-level cities (Fig. 2.1).

To account for the potential of accessing green spaces beyond the delineated boundaries, an extension of the study area for each city was established, which

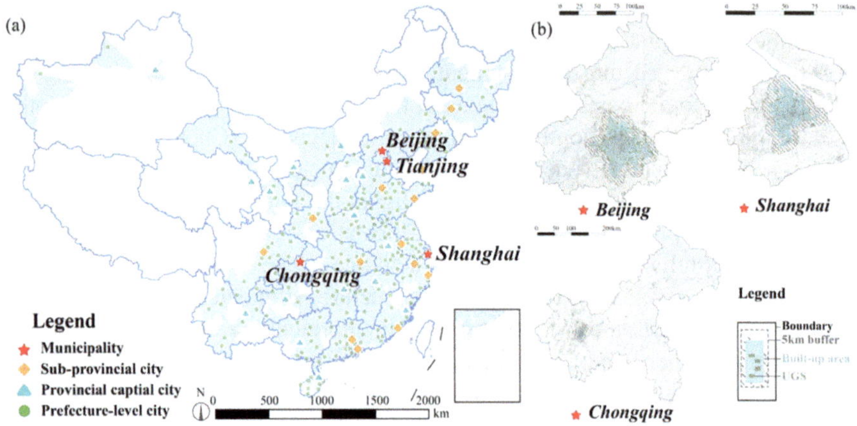

Fig. 2.1 National-scale distribution of sample cities

encompassed a buffer zone extending 5 km beyond the urban built environment. The data on residential communities were derived from SouFun Holdings (http://www.soufun.com.cn) in 2020, encompassing information on 301,243 residential communities nationwide, including details such as location, average price, green space rate, and household count. However, some communities exhibited incomplete data on green space rate and household numbers. To complete this information, we employed web crawling technology via Loupan (https://www.loupan.com). After data cleaning, 106,841 residential districts were retained for the study. We acquired Landsat cloud-removed imagery from August 2020 to compute the NDVI (NDVI), thereby avoiding computational biases. The resource endowment metric for residential districts is calculated as follows:

$$Q_i = \mathrm{NDVI}_{i0} + \mathrm{NDVI}_{i1} + \mathrm{NDVI}_{i2}$$

where NDVI_{ij} represents the average NDVI value within a square buffer extending j pixels outward from the central pixel of residential area i (as derived in the preceding subsection). For instance, denotes the mean NDVI within a 3 × 3-pixel grid centered on the pixel where residential area i is located, formed by extending 1 pixel outward. The cumulative use of multi-scale buffer values mitigates the Modifiable Areal Unit Problem (MAUP) in quantifying resource endowments.

We identified and consolidated the presence of PGS within urban built-up areas using Points of Interest (POIs) and Area of Interest (AOIs) from three reputable sources. POIs were sourced from categories encompassing tourist attractions and natural features on Baidu Maps, as well as the scenic attractions category on Amap. AOIs were extracted from boundary data pertaining to green spaces from OpenStreetMap, Baidu Maps, and Amap. By cross-referencing and validating these datasets against each other, we ensured the integration of comprehensive and accurate green space boundary data. Based on the green space boundary data, we calculated navigation routes utilizing the Amap API, which can facilitate the calculation of routes and travel times from communities to nearby PGS, thereby assessing the accessibility of each community to PGS.

We measure intra-urban green inequality across residential districts using a modified generalized Gini index (Biró and Néda 2020; O'Donnell et al. 2016). Unlike conventional algorithms, we define the horizontal axis as the average housing price of residential districts. After sorting districts from lowest to highest price, the vertical axis represents the cumulative green exposure level. This approach more effectively captures the relationship between socioeconomic disparity and green accessibility. The Gini index ranges from 0 to 1, and a lower Gini index means people in different SES can assess a more even amount of green space.

The distribution of UGS correlates with a series of socioeconomic and ecogeographic factors. We constructed a multivariate regression model with the Gini index as the dependent variable and selected socioeconomic variables that were significantly correlated with the Gini index as independent variables, including total population, population growth rate, GDP per capita, the proportion of secondary industry in GDP, the proportion of tertiary industry in GDP, electricity consumption, and average housing price. Population data from the 2010 and 2020 Chinese National Population Censuses, and GDP-related metrics from the 2021 China City Statistical Yearbook. To account for economic disparities driven by geographic heterogeneity, cities were categorized into four economic regions (Eastern, Central, Western, and Northeastern China) with dummy variables. Ecogeographic city factors, including temperature, average green space rate, average NDVI, and average PGS accessibility, were included as control variables. Precipitation and humidity were disregarded and omitted from the analysis because of their non-correlation with the dependent variable. Additionally, exploratory regression analyses were conducted for each city, treating housing price as the independent variable while separately modeling greening rate, NDVI, and accessibility as dependent variables.

2.1.3 Characterization of Urban Green Spaces and Their Equity at the National Scale

Figure 2.2 illustrates the contribution of greening rate, NDVI, and accessibility to overall green exposure across different cities. Results reveal significant spatial heterogeneity in green space exposure among cities and regions, with higher-tier cities typically exhibiting superior accessibility. Socioeconomic factors negatively impact at least one dimension of green equity in all studied cities. In most urban areas, green space accessibility is notably compromised—particularly in megacities such as Beijing, Shanghai, and Tianjin. Emerging metropolises (e.g., Hangzhou, Suzhou, Shenzhen) show significant effects across all three dimensions of green exposure. Substantial inter-city disparities in Gini index are observed (Fig. 2.3). Eastern cities demonstrate higher Gini index across all three metrics compared to other regions,

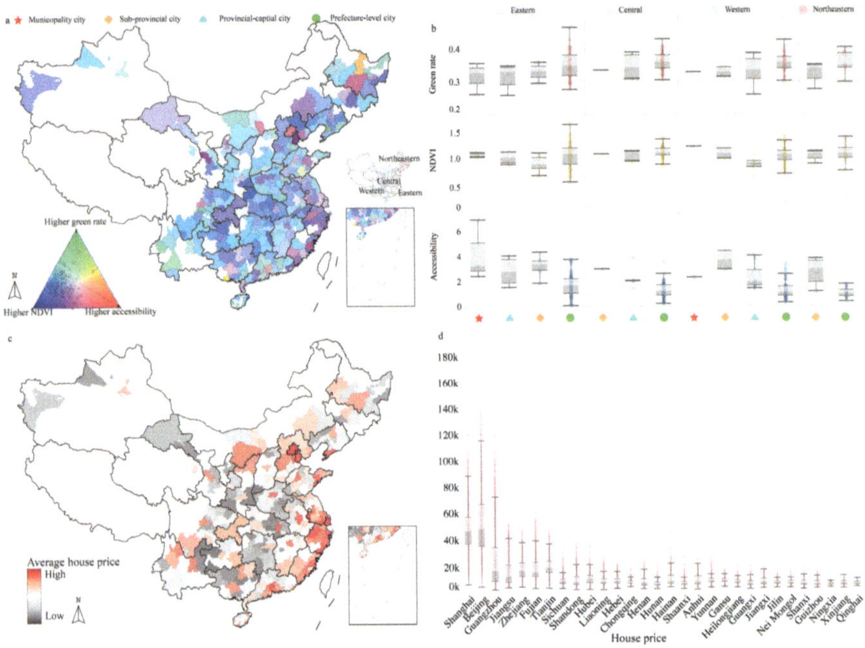

Fig. 2.2 (**a**) Contributions of green space ratio, NDVI, and green accessibility across 263 Chinese cities; (**b**) three metrics in cities of different economic policy zones; (**c**) average housing prices in 263 Chinese cities; (**d**) inter-provincial housing price comparison

yet those with elevated administrative status (e.g., provincial capitals) tend to exhibit lower inequality. In western China, provincial capitals and prefecture-level cities display nearly equivalent levels of green equity. Conversely, in northeastern China, prefecture-level cities outperform provincial capitals in equity performance. Local spatial autocorrelation analyses indicate that green inequity driven by residential segregation and resource endowments clusters in southern cities, whereas accessibility inequality is more pronounced in eastern coastal cities. Cities with low Gini index are concentrated in the northeastern and southwestern urban agglomerations of China, providing more equitable public green space accessibility.

A comparative analysis of green equity levels across three dimensions is presented (Fig. 2.4). In Shandong Peninsula, Guangxi, and Jiangxi, the three metrics demonstrate high consistency. However, in the Yangtze River Delta, Pearl River Delta, and Hainan, disparities in equity reflected by the three metrics are observed. We further employed K-means clustering to analyze the convergence and divergence of these three green equity dimensions. The clustering results (Fig. 2.5) reveal that metropolitan areas including the Yangtze River Delta, Pearl River Delta, Beijing-Tianjin-Hebei, and Shandong Peninsula are all classified into Cluster 1, where most prefecture-level cities exhibit pronounced green inequity. Cluster 4 primarily represents cities in Hubei, characterized by a Gini index of accessibility significantly higher than those of residential segregation and resource endowment.

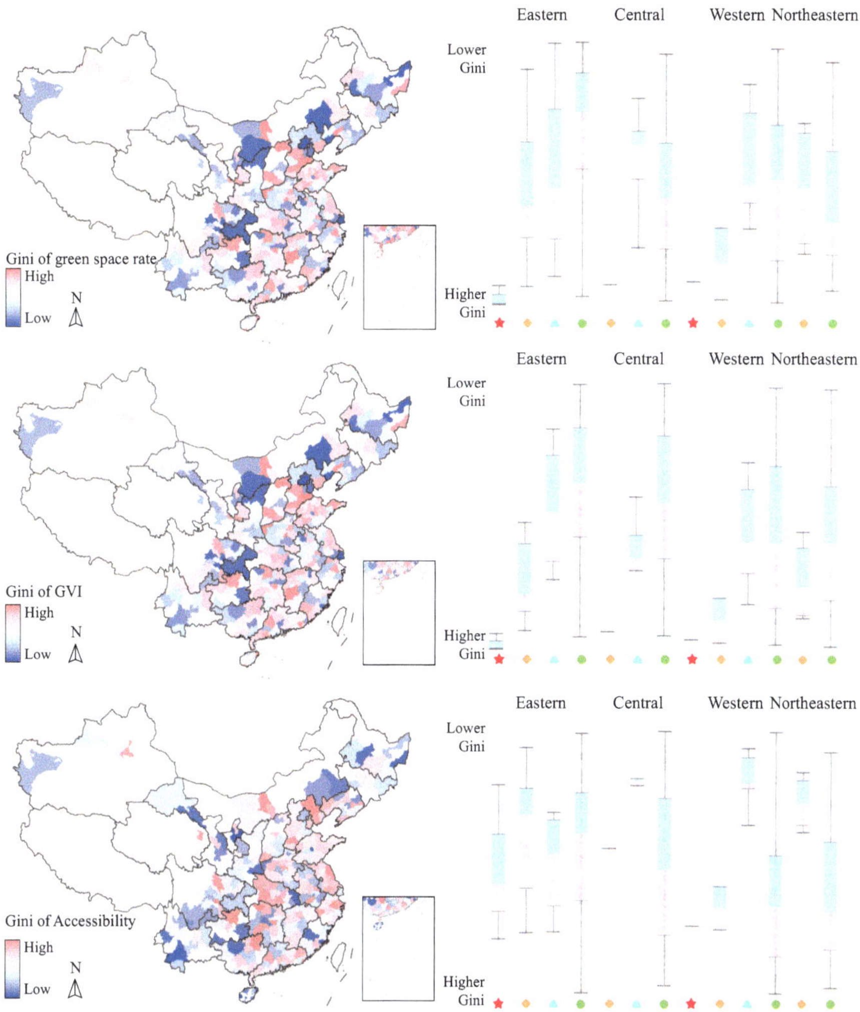

Fig. 2.3 Green space equity and spatial heterogeneity in Chinese cities

Cluster 5 consists of cities where all three metrics consistently indicate high green equity, predominantly located in the western region.

Multivariate regression analysis reveals that economically developed megacities exhibit higher equity in residential segregation and resource endowment, but paradoxically demonstrate greater inequality in park accessibility. City administrative hierarchy and economic regional divisions significantly influence green equity levels, with municipalities directly under the central government and western cities showing relatively higher equity. Further analysis indicates that although residential segregation and resource endowment are similarly affected by socioeconomic factors—per capita GDP shows a negative correlation with the Gini index (suggesting

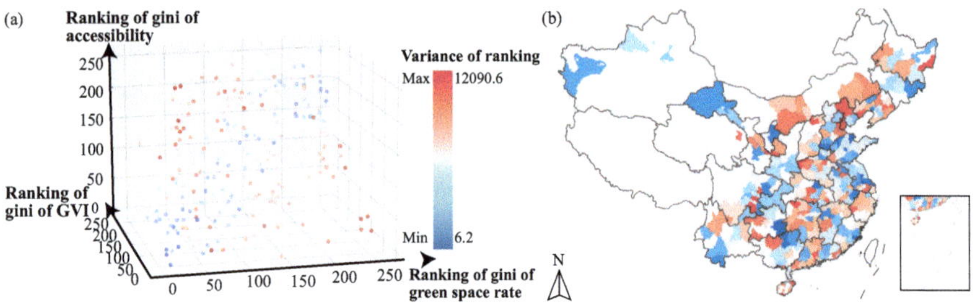

Fig. 2.4 (**a**) 3D visualization of Gini index ranking; (**b**) spatial visualization of Gini index variance

higher equity in affluent cities)—accessibility presents a contrasting pattern: per

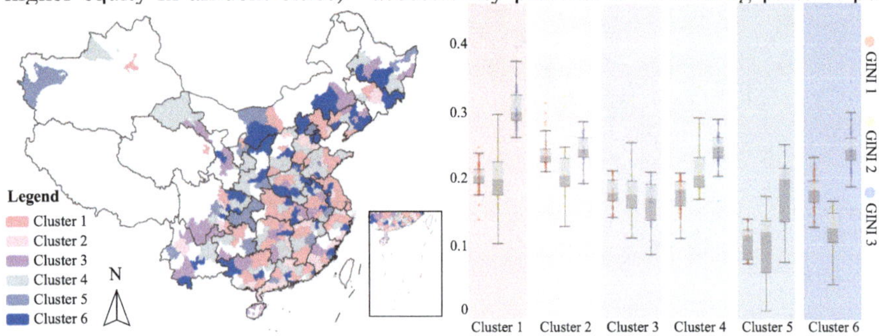

Fig. 2.5 Cluster analysis results

capita GDP and population size positively correlate with accessibility-related Gini index. Additionally, increased proportions of secondary and tertiary industries exacerbate inequality across all three dimensions. Significant disparities emerge between city tiers and economic regions, with municipalities achieving the highest green equity levels, eastern cities the lowest, and western regions maintaining moderate equity.

Regression results for individual cities reveal that in most urban areas, residential segregation and resource endowment show no significant correlation with residents' socioeconomic attributes—a finding contradicting theoretical expectations. This anomaly may stem from targeted governmental interventions (e.g., old neighborhood renovation programs) that restrict high-income groups' access to privileged green exposure. Conversely, accessibility equity is neglected due to China's reliance on land finance, resulting in substantially higher public green space access for socioeconomically advantaged groups, thereby exacerbating disadvantages for marginalized populations. Nationally, residential segregation and resource endowment exhibit similar patterns of green inequity, particularly in southern cities where low equity levels prevail—likely linked to green space scarcity driven by post-1980s manufacturing growth and migrant influx. Regions benefiting from economic reforms (e.g., Yangtze River Delta, Beijing-Tianjin-Hebei, Shandong Peninsula) cluster with southern cities in high-inequity categories. Western cities demonstrate better residential segregation equity, attributable to universal compliance with minimum greening policies despite harsh climates and high maintenance costs. Northeastern cities excel in resource endowment equity, leveraging urban shrinkage from population decline and deindustrialization, alongside the Three-North Shelterbelt Program to advance higher-quality green infrastructure. Accessibility equity, however, diverges markedly: eastern coastal cities exhibit severe inequity despite superior administration and economic resources, where green space loss and excessive built-up expansion degrade ecology and create park accessibility disparities—particularly in urban–rural fringes excluding low-income residents from mandated 15-minute green circles. Consequently, China's socialist green space planning must prioritize marginalized groups to ensure equitable access.

The study reveals that per capita GDP exerts a negative impact on the Gini index of residential segregation and resource endowment, while population growth rate shows a positive effect. This indicates that economic development facilitates green equity, yet population influx and economic expansion exacerbate inequity in green space accessibility. GDP growth provides local governments with increased fiscal revenue, incentivizing them to prioritize greening coverage, ensure uniform vegetation provision, and enhance residential segregation equity. However, the green space expansion driven by GDP growth and population surge predominantly occurs in low-cost urban peripheries, with governments reluctant to augment green infrastructure in city cores, thereby intensifying accessibility disparities. Significant differences emerge in green equity levels across cities of varying administrative tiers: higher-tier cities exhibit superior green equality, reflecting their heightened focus on addressing inequality stemming from wealth polarization. Key drivers for equitable green space provision in these cities include improving resident well-being, sustainable urban development imperatives, and inter-city competitiveness. Notably, eastern China—the most economically advanced region—demonstrates only marginally better green equity than the northeast, potentially due to eastern governments prioritizing income generation over ecological considerations.

The increasing proportions of secondary and tertiary industries exhibit negative impacts on equity levels of residential segregation, resource endowment, and accessibility, with secondary industry exerting particularly pronounced effects. This phenomenon correlates with China's economic-oriented industrial structure over the past four decades, wherein local governments allocated land at subsidized prices to industrialists to accelerate industrialization, reserving only marginal lands unsuitable for construction as parks—a practice that institutionalized unequal green exposure. While industrial zones occupy vast natural green spaces, their impacts on intra-community green infrastructure remain limited, yet significantly distort accessibility equity. Cities with high tertiary industry concentration tend to develop dense commercial enclaves in peripheries, where affluent populations monopolize ancillary green amenities, thereby exacerbating inequity. Paradoxically, the "National Garden City" designation—a policy incentive—has been awarded not to cities with equitable green distribution but to those with worse equity performance, revealing systemic flaws in China's greening policies: evaluation criteria prioritize quantitative metrics (e.g., per capita park area, green coverage rate) while neglecting spatial justice considerations, potentially hindering cities from improving green space equality.

These findings provide urban policymakers with an evidence-based understanding of urban green space inequity. Addressing this challenge necessitates establishing robust public participation mechanisms and adopting the Gini index as a spatially explicit method to evaluate planning rationality. Furthermore, this research bridges critical gaps in existing literature regarding understudied small cities and Global South urban contexts, offering actionable insights for developing inclusive cities in South America and Africa.

2.2 Optimizing Urban Green Space Equity Through Explainable Machine Learning

Over the past few decades, research assessing green justice has been on the rise. A large body of literature has examined UGS equality within cities by investigating spatial patterns and their linkage to multiple benefits. These studies can be roughly divided into two categories based on scale. One category evaluates the accessibility or availability of green spaces for urban residents with different socioeconomic attributes within a single city and then further discerns the level of green justice (Chen et al. 2021; Comber et al. 2008). Within these studies, definitions of UGS vary, as do the measurement methods and demographic subgroup indicators, leading to diverse conclusions for individual cities. To overcome this inconsistency, another line of research makes cross-city comparisons of green justice levels using consistent indicators and measurement methods (Chen and Wang 2013; Nesbitt et al. 2019; Zhu and Zhang 2008). Some studies aim to understand regional differences and the driving factors of green justice. Cross-city studies suggest that green justice levels may be determined by a combination of interrelated factors, including socioeconomic, biophysical, and historical influences. Such factors are complex and profound, require long-term societal processes to change, and do not provide a clear approach to enhancing urban green justice levels.

However, the spatial characteristics of UGS themselves and their impact on green justice have been largely overlooked. The morphology, size, boundaries, layout, and proximity of UGS directly affect residents' accessibility and availability of UGS, subsequently influencing green justice levels (Gu et al. 2024). Yet, how specific green space spatial configurations can promote green justice has not been systematically discussed. Moreover, the associations between UGS spatial characteristics and green justice might be non-linear, and empirical evidence on this topic remains sparse. Many studies on factors driving green justice presume linear or predefined patterns, neglecting potential non-linear relationships and interaction effects between variables. In this sense, they are likely to overestimate or underestimate the effects, and the combination of two variables may amplify or shrink those effects.

Addressing these literature gaps requires quantitatively examining the associations between UGS attributes and green justice in Chinese cities, with specific emphasis on non-linear characteristics and interactive effects. An advanced machine learning method, the Gradient Boosting Decision Tree (GBDT), can conveniently deal with non-linearity and synergism. Shapley additive explanations (SHAP), grounded in game theory, should be adopted to locally interpret the influence of UGS characteristics, thereby offering granular insights to advance green justice.

2.2.1 Spatial Characteristics of UGS

The spatial characteristics of urban green spaces might play a pivotal role in shaping UGS equity. UGS characteristics, as external manifestations of socioeconomic, biophysical, and historical factors in the urban environment, might directly influence residents' opportunities to access and experience green spaces, determining the accessibility to UGS. To explore the relationship between these characteristics and green justice, we theoretically constructed a series of green space characteristics that might impact green justice, encompassing both individual (measure for each UGS, including planar and vertical morphology) and collective green space characteristics (measure for UGS in the entire city). We reviewed existing supporting literature, ensuring that each indicator holds profound significance and relevance in ecological, social, and statistical contexts.

Spatial characteristics of urban green spaces are considered pivotal in shaping UGS equity. Such characteristics, viewed as external manifestations of socioeconomic, biophysical, and historical influences within the urban environment, are understood to determine opportunities for accessing and experiencing green spaces, thereby affecting UGS accessibility. To examine relationships between these characteristics and green justice, a theoretical framework of green space attributes was developed, encompassing both individual-level metrics for each UGS (including planar and vertical morphology) and aggregate-level metrics for the entire urban area. Existing literature was reviewed to ensure that all selected indicators possess substantial ecological, social, and statistical relevance.

The selected planar-morphology metrics are used to capture multiple dimensions of urban green-space patches—size, proportional coverage, boundary complexity, shape irregularity, and internal heterogeneity. Average green-space area (GAA) and green-space area proportion (GAP) quantify typical patch size and overall coverage; their coefficients of variation (GAVC) reveal heterogeneity in patch sizes. Small-patch metrics (SGP and SGAP) focus on the prevalence and relative area of patches below a study-wide size threshold. Total edge length (TE) and mean perimeter (GMP), together with the perimeter variation coefficient (PVC), describe boundary complexity and its variability across patches. Shape-complexity indicators—including the landscape shape index (LSI), mean fractal dimension (MFD), and area-weighted fractal dimension (AWGFD)—assess overall shape irregularity, while the fractal-dimension variation coefficient (GFDVC) reflects diversity in patch complexity. Finally, compactness measures (mean compact ratio, MCR; area-weighted compactness ratio, AWGSCR) and their variation coefficient (GSCRVC) evaluate how closely patches approximate simple shapes and the degree to which this compactness varies across the urban landscape. Together, these indicators provide a comprehensive characterization of planar morphology relevant to equitable access and experiential quality of urban green spaces.

The vertical morphology of green spaces reflects the vegetation abundance, ecological features, and health status of green spaces in cities. This mainly includes the Mean Normalized Difference Vegetation Index (MNDVI) of UGS, the Maximum NDVI (MAXNDVI), and the NDVI Variation Coefficient (NDVIVC). Firstly, both MNDVI and MAXNDVI collectively indicate the overall ecological health of urban green spaces, where higher values typically imply that the vegetation in urban green spaces is in good health and growth condition. From a social viewpoint, a higher MNDVI value also suggests that green spaces in the city are capable of providing leisure and recreational areas for a larger portion of urban residents, contributing to achieving green justice. Secondly, the NDVIVC offers information about the unevenness of the vegetation conditions in UGS. These indicators reflect the competency and efficacy of urban administrators in managing green spaces. This indirect indication could imply the presence of substantial green justice issues within the city's management practices.

The collective morphology of green spaces reveals the organization of green spaces in urban areas, offering a valuable perspective for evaluating the associations between the structure and distribution of the urban ecological network and green justice. Firstly, indicators related to the collective structure, including the Euclidean Nearest Neighbor distance (ENN) and Contagion (CONTAG), primarily reflect the connectivity and proximity of UGS. A highly interconnected green space network can promote the interaction of people, goods, and information flow among the various green spaces, enhancing the overall utilization efficiency of the urban green space system and, to an extent, increasing accessibility equity. Secondly, indicators for group distribution include eccentricity degree and dispersion degree, which reveal the spatial distribution characteristics of UGS within the city. These indicators can reflect whether the distribution of green spaces is even or if there are tendencies of concentration or marginalization, consequently influencing the extent of enjoyment of green spaces by communities and groups located in different areas of the city.

In addition, socioeconomic indicators of the city were also incorporated as control variables. To clarify the data sources for each indicator, a chord diagram was created and is presented in Fig. 2.6.

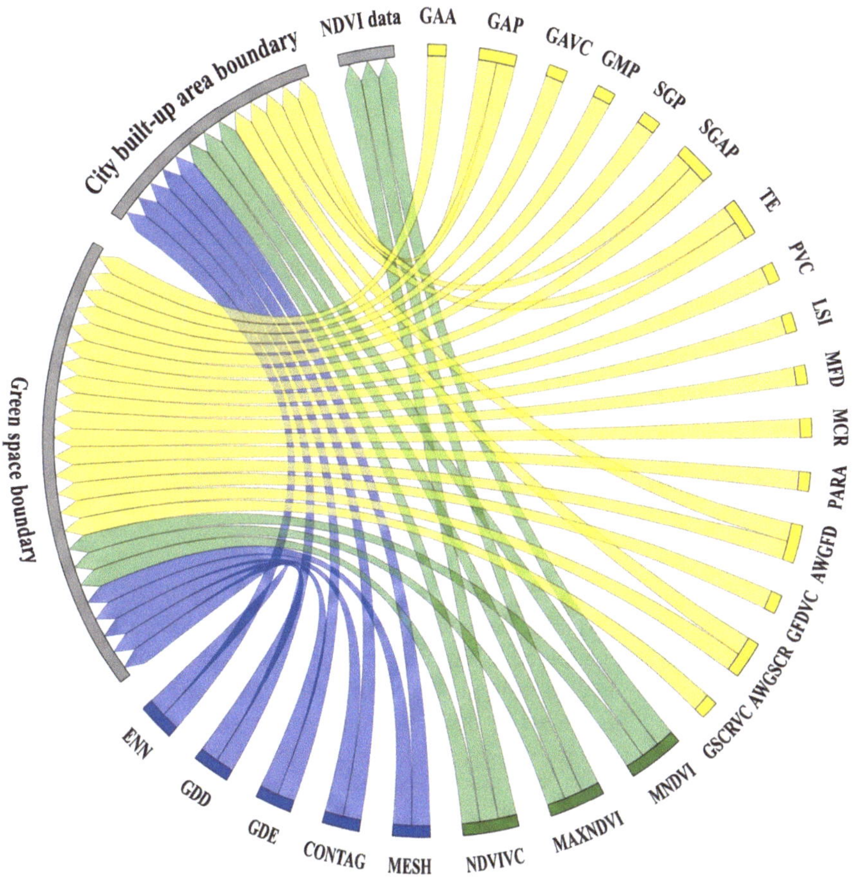

Fig. 2.6 Chord diagram illustrating the data utilized for each indicator

2.2.2 Modeling and Interpretation Method

Machine learning methods are preferred for the following reasons. First, sample-size requirements are relatively low, making such methods well suited to cross-city analyses with modest datasets (Hastie et al. 2001). Second, multicollinearity is handled robustly, so that interdependent landscape indicators do not compromise model accuracy (Friedman 2001). Third, non-linear relationships among variables can be uncovered directly from data without prespecified functional forms. Fourth, interaction effects among predictors are captured automatically, avoiding the need to manually specify interaction terms.

On account of these strengths, LightGBM (Light Gradient Boosting Machine) was selected as the modeling approach (Ke et al. 2017). As an optimized variant of gradient boosting decision trees, it applies a leaf-wise growth strategy—allocating resources to the most loss-reducing splits—and employs histogram-based binning and sparse-feature optimizations to accelerate training and reduce memory use (Shi et al. 2019). Its interpretability is augmented through both global metrics (feature-importance rankings) and local attributions derived from Shapley-value explanations (SHAP). Primary and interaction effects can be quantified via SHAP interaction values, revealing how pairs of features jointly influence predictions (Kim and Kim 2022).

Model development was conducted in a Python 3.10 environment using LightGBM together with Scikit-learn and Scikit-optimize. Fivefold cross-validation and Bayesian optimization were applied for hyperparameter tuning to maximize generalization performance. Urban green-space indicators served as predictors, and the Gini index of green-space accessibility served as the target variable. Model performance was evaluated on both training and test sets, demonstrating high accuracy and reliability in predicting green justice levels.

2.2.3 *Evaluating the Impact of UGS Characteristics on UGS Equity*

The results indicate varied associations between different types of UGS indicators and green justice levels. Among them, the green space collective characteristics and planar morphology are identified as those with the strongest associations based on the global importance derived from the cumulative SHAP values (see Fig. 2.7). On the left side of Fig. 2.7 is the SHAP decision plot, where each line represents a city, and the colors denote the equity of green space accessibility (blue being more equitable and red less equitable). Starting from the baseline at the bottom, the contributions of each indicator cumulatively add up to the predicted value. GDD is identified as the dominant influencing factor, followed closely by the proportion of SGP and the LSI. This highlights the pivotal role that the distribution and configuration of green spaces play in green justice. In contrast, we found that the vertical morphology indicators of green space have smaller associations with green justice, implying that the internal ecological quality of the green space might not significantly affect green justice.

On the right side of Fig. 2.7 is the SHAP bee swarm plot, with each dot corresponding to a city. A dot's position on the x-axis shows the indicator's association with the city's Gini index. Positioned to the left of the central axis indicates a negative association, and to the right indicates a positive association. When multiple dots fall on the same point on the x-axis, they stack up to display density, with color representing the value of the indicator. For instance, the dispersion degree of GDD has a long blue tail on the left and a short red tail on the right. This suggests that a low GDD might decrease the Gini index, i.e., enhance the level of green justice, but a high GDD is unlikely to improve the green justice level to the same extent. The Shapley values for cities with high GDD (red) exceed those with low GDD (blue), implying a positive association between GDD and the Gini index. Overall, these results reveal that the associations between urban green space spatial characteristics and green justice are multifaceted and complex. Different features have varying

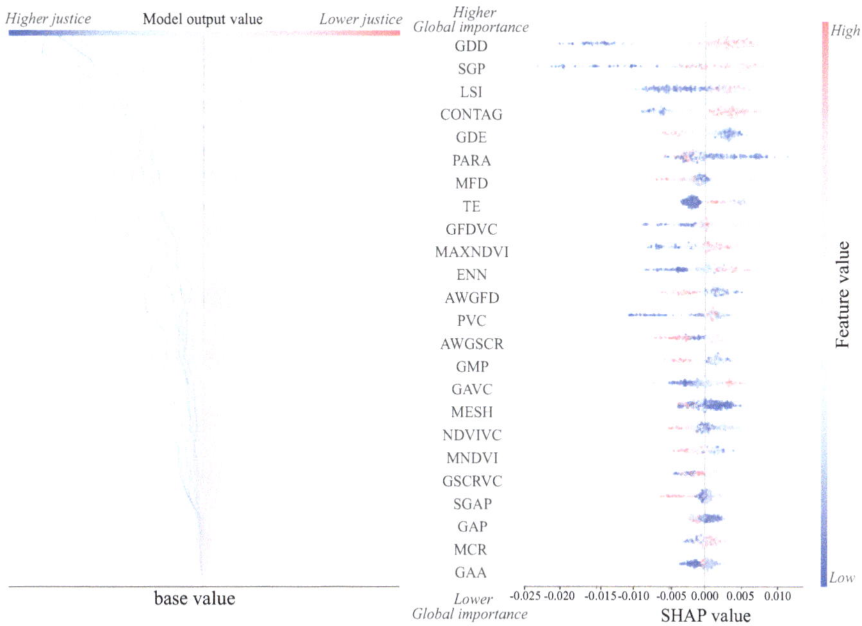

Fig. 2.7 Shapley value and importance rank of independent variables

associations, providing important directions for our further exploration of green justice.

Local dependence plots are utilized to reveal the non-linear relationship between green space characteristics and green justice level. As illustrated in Fig. 2.8, each local dependence plot corresponds to a green space spatial characteristic indicator, demonstrating how the variable alters the green justice level for each city. Each point in the figure corresponds to a city.

Overall, Fig. 2.8 validates our non-linear relationship hypothesis, showing that the effect of UGS indicators on green justice levels varies as the value range changes, with additional threshold effects. The local effect of GDD, which ranks highest in importance, first declines and then rises, correlating positively with justice levels. When GDD values exceed 0.5, its local effect escalates to a positive value. This implies that initially, as the green space distribution becomes more dispersed, the level of green justice increases. However, excessively dispersed green spaces result in a decline in green justice levels. Furthermore, we observed that the key indicator of green space proximity, ENN, has an increasing SHAP value trend. When ENN exceeds 450, its SHAP value turns positive, suggesting that an average green space gap exceeding 450 m likely has a negative association with UGS's just accessibility.

Within the green space planar morphology indicators, the proportion of small green spaces is the most notable. It aids in enhancing the green justice level when constituting <80 % (see Fig. 2.8). However, if a city is entirely composed of small green spaces, the justice level decreases. The increase in LSI values, indicative of

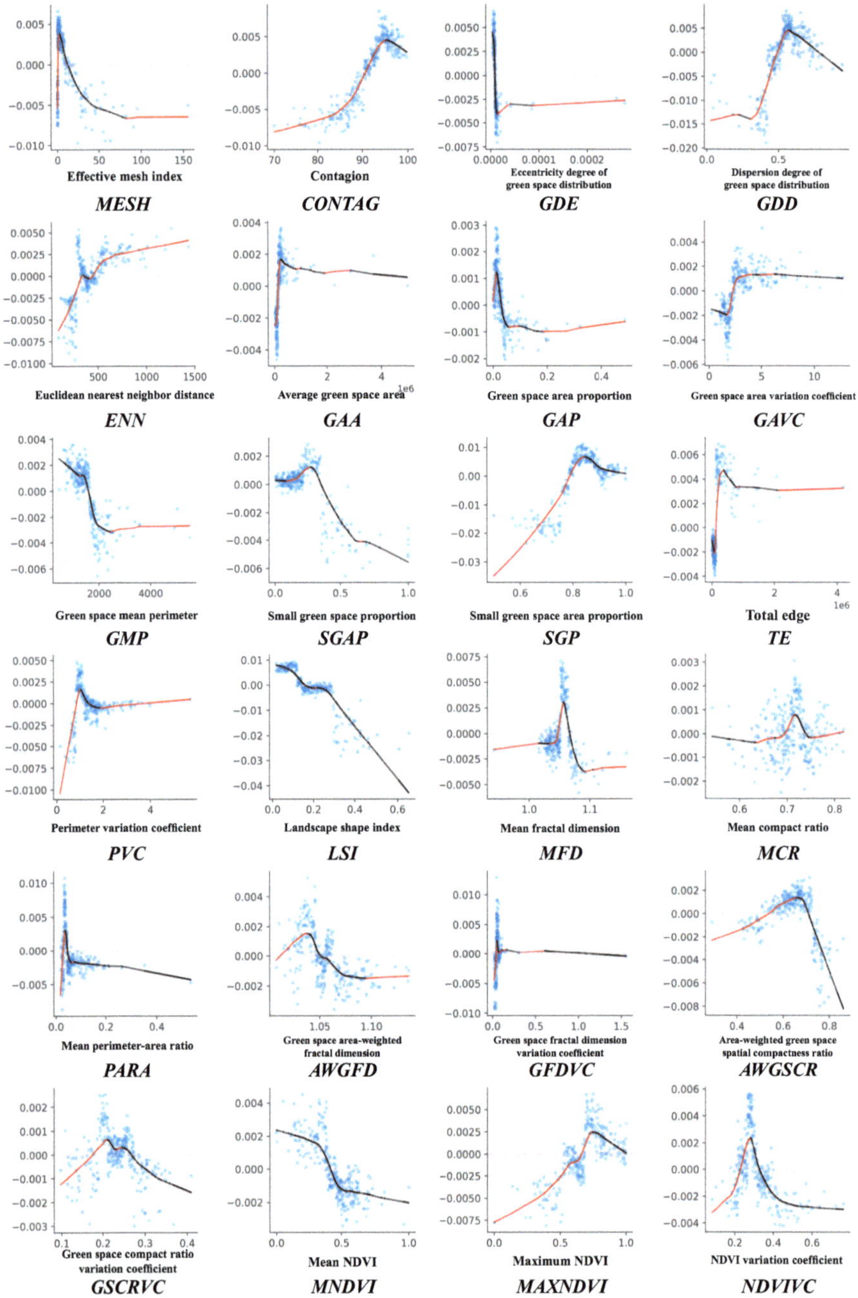

Fig. 2.8 Local dependence plots (Note: A positive value means an increase in the Gini index, indicating a decrease in the level of green justice. Negative values imply the opposite. If, within a certain range, the Gini index rises with an increase in the value of UGS indicators, the trend line is red; otherwise, it is black.)

the irregularity of green space (i.e., deviating more from square or circular shapes), consistently enhances green justice levels. Similarly, a positive correlation exists between the PARA and justice levels, suggesting that the longer and more complex the boundary, the higher the justice level. When the GMP exceeds 1600 m, justice levels rise. In terms of vertical morphology indicators, both MNDVI and the NDVIVC correlate positively with the green justice level. This indicates that cities with greater vegetation abundance and green space heterogeneity typically have higher green justice levels.

The local effect of an indicator is composed of its main local effect and its local interaction effects with other indicators. The local interaction effects of each indicator with others were evaluated (based on the average absolute Shapley interaction values), selecting and visualizing the indicator that has the strongest interaction effect. Figure 2.9 depicts the effects of these highly interactive indicator pairs, with each pair in the subplots illustrating how they are associated with green justice. Every dot represents a city, with its color denoting the other variable that has the strongest interaction effects with it.

Certain combinations of indicators have been found to be significantly associated with green justice, thereby filling gaps in prior research. Within collective morphology metrics, notable interaction effects occur between ENN and SGP: when ENN < 500 m and SGP < 75%, or when ENN is 500–1000 m and SGP is 75–85%, green justice should be markedly improved.

Among planar morphology metrics, the interaction between GAP and SGAP should play a key role in enhancing green justice. In contexts of low GAP, an increase in SGAP should boost green justice levels. Furthermore, under high SGAP conditions, greater CONTAG should correspond to higher green justice levels. The relationships involving GMP and SGP, as well as PARA and SGP, are more complex: combinations of GMP < 1600 m with SGP > 45%, or GMP > 1600 m with SGP < 20%, should enhance green justice; similarly, when both PARA and SGP are either low or high, green justice should improve more readily.

Within vertical morphology metrics, a significant interaction effect should exist between MNDVI and AWGFD: when MNDVI < 0.45, AWGFD should be minimized to achieve notable improvements in green justice. In summary, optimal enhancement of green justice should be guided by careful consideration of indicators such as distribution (ENN, GAP, SGP), proximity, spatial compactness (CONTAG, MCR), and shape complexity (GMP, PARA, AWGFD), to ensure the most effective allocation of UGS resources.

2.2.4 *Pathways for Optimizing Urban Green Space Equity*

By revealing the local effects of UGS indicators in each city, valuable insights can be provided to enhance the green justice level. A crucial approach to achieving this objective is to reveal similar local effect patterns among cities, thereby pointing out targeted planning and strategies for different city types. Ultimately, this approach

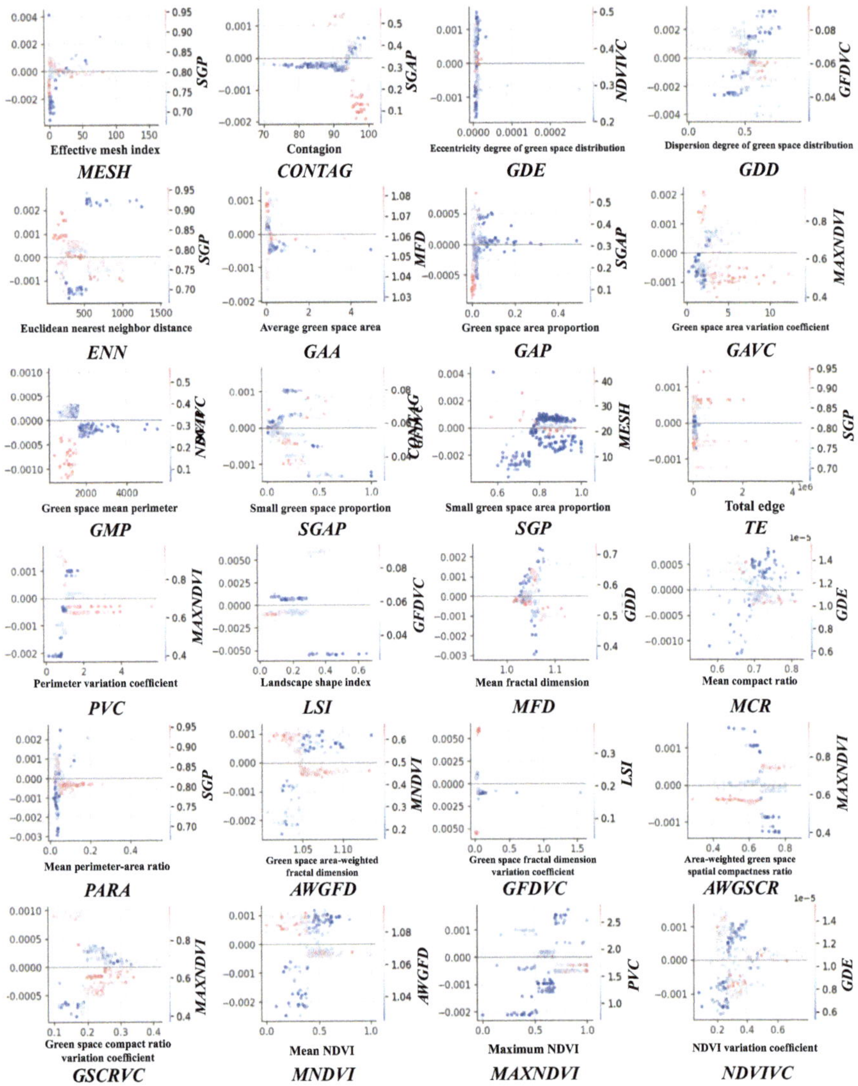

Fig. 2.9 Interaction effects among UGS indicators (Note: In each subplot, the horizontal axis represents the value of the variable below, and the color of each point represents the value of the indicator that has the strongest interaction with the variable below. The higher the vertical axis, the worse the urban green justice level caused by the interaction between the two variables. The lower the horizontal axis, the better the green justice level resulting from the interaction.)

aims to suggest both immediate and long-term strategies for various city types, accommodating their unique histories and growth trajectories. As shown in Fig. 2.10a, cities are sorted along the x-axis based on similarity, with corresponding Shapley values stacked along the y-axis. Positive Shapley values (in red) decrease the green justice level, whereas negative Shapley values (in blue bars) increase it. When a distance threshold of 0.12 is applied, four clusters stand out, containing 63,

40, 71, and 88 cities, respectively. Cluster 3 has the lowest green justice level, while Cluster 1 has the highest.

It is evident from Fig. 2.10b that clusters based on local-effect similarity exhibit clear spatial concentrations. Cluster 1 is concentrated in northern frontier regions, where cities tend to have limited scale and economic development. Cluster 2 is located in central-southern regions, characterized by low scale and economic advancement. Cluster 3 occupies eastern coastal regions, encompassing most first-tier cities with large areas and high GDP levels. Cluster 4 covers central regions, featuring cities with relatively higher average area and GDP. Further analysis revealed that GDD and SGP were the indicators most negatively associated with green justice across all clusters. Consequently, GDD and SGP ought to be prioritized regardless of city size or economic status, although their optimal directions differ: an increase in GDD is recommended for Cluster 3, whereas a reduction is advised for Clusters 1, 2, and 4; conversely, SGP should be increased in Clusters 1, 2, and 4 but reduced in Cluster 3.

Beijing, Shanghai, and Guangzhou from Cluster 3 and the highest-GJ cities from Cluster 1 (Wuzhong, Puer, and Dongfang) were selected as case studies to examine indicator–justice associations in representative contexts (Fig. 2.10c). Shapley-value analysis revealed both commonalities and differences across clusters. In Beijing, Shanghai, and Guangzhou, ENN was positively associated with green justice, and PARA values were relatively low, promoting access. Despite SGP exceeding 90 %, a negative association with green justice was observed. MESH values in Beijing and Guangzhou were high—indicating continuous layouts and positive justice effects—whereas Shanghai's low MESH (2.84) reflected fragmentation and a negative association. NDVIVC was lowest in Shanghai (0.30), suggesting uniform cover but potential equity drawbacks.

Shapley-value observations for Wuzhong, Puer, and Dongfang showed similar patterns: GDD and PARA exhibited positive effects on green justice, and consistently low values of both indicated that concentrated, well-organized UGS enhances equity. ENN was positively linked to justice in Wuzhong and Puer but negatively in Dongfang. SGAP showed the strongest association with justice in Puer, while its influence was minor in Wuzhong and Dongfang.

Amid rapid urbanization—especially in developing countries—concern for the equitable distribution of urban green space continues to intensify. Existing cross-city investigations have largely concentrated on socioeconomic determinants of green justice yet have seldom incorporated landscape metrics that directly condition UGS accessibility. Because each city's physical form, land-use pattern, and broader socioeconomic setting differ, the principal drivers of green justice, and the strategies required to improve it, likewise vary. A systematic exploration of the linkage between key green space metrics and equity therefore ought to guide policy formulation.

Evidence identifies four metrics with the strongest associations to green justice. GDD emerges as the single most influential factor. Although higher dispersion permits residents in diverse locations to reach nearby UGS, an optimal range is apparent: when GDD lies between 0.25 and 0.50, accessibility is maximized; values above this threshold create underserved voids, while excessively low dispersion

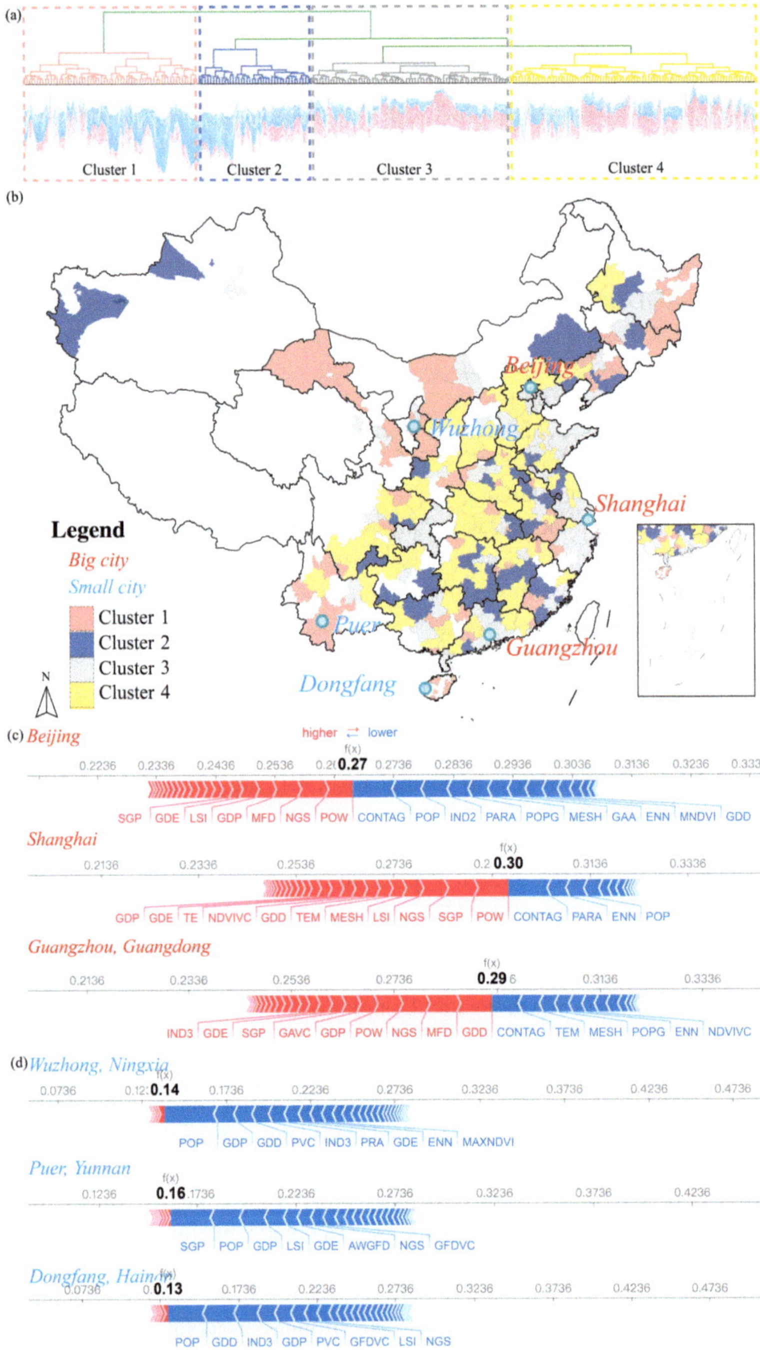

Fig. 2.10 (**a**) Results of clustering; (**b**) clusters corresponding to cities; (**c**) local effects of the case big cities; (**d**) local effects of the case small cities

concentrates amenities in limited districts. Urban plans should therefore calibrate GDD to fall within this intermediate band, ensuring that residents across income strata obtain reasonable local coverage without suffering the deficits caused by either fragmentation or over-aggregation.

SGP ranks second in importance. Large parks frequently raise adjacent land values and living costs, risking exclusion of vulnerable groups, whereas numerous micro-parks may fail to supply sufficient usable area. Equity improves only when the cumulative area of small parks meets local demand. Thus, cities characterized by scant overall green coverage should enlarge aggregate SGP by adding not merely more—but also larger—small parks; by contrast, cities already rich in greenery should prioritize increasing the count of small parks to refine fine-grained access.

ENN provides the third major influence. A mean inter-park distance below 450 m—closely matching the canonical 500-m neighborhood-walkability radius—correlates with markedly higher equity. Once ENN exceeds this threshold, benefits decline. Urban design should therefore aim to maintain average park spacing beneath the 450-m benchmark, thereby reinforcing both social and ecological sustainability.

Finally, boundary geometry strongly shapes equitable outcomes. Higher perimeter-to-area ratios and greater boundary complexity lengthen edge zones—the interface where most social and ecological exchanges occur—and consequently raise justice levels. Green space design should, wherever practicable, favor compact yet irregular outlines that expand accessible perimeters without excessive land take.

Taken together, calibrated dispersion, adequately extensive small-park systems, sub-450 m spacing, and complex but compact boundaries should be pursued to optimize urban green space equity.

Rapid urbanization has intensified public concern over the equitable allocation of urban green space, and recent analysis indicates that equity outcomes hinge on several key landscape metrics. GDD exerts the strongest influence: equity improves as dispersion rises only to an optimum band of roughly 0.25–0.50, beyond which voids emerge and justice declines. Large, polycentric cities—especially those assigned to Cluster 3—tend to display highly dispersed residential patterns, a consequence of migrant inflows; their green networks therefore ought to be spread even more widely to ensure that every neighborhood retains convenient access. By contrast, single-centered, smaller cities concentrate housing within older districts; in these settings, an overly high GDD would squander resources, so a tighter pattern should be maintained. Attention must be paid to the presence of gated compounds, whose impermeable boundaries frequently nullify the theoretical benefits of higher GDD; in such cases, internal greening and permeable edge retrofits should be prioritized.

SGP supplies the second-most potent leverage point. In the largest conurbations the count of pocket parks already exceeds demand, and further subdivision would dilute functional value; efforts should instead enlarge existing sites and enrich their programming. Smaller cities, conversely, suffer from sparse SGP; modest, widely scattered interventions—mini-greenways, community gardens and similar devices—can raise coverage without provoking land-price inflation or green gentrification. The decisive variable is not the sheer number of parks but the aggregate area available for daily use.

Proximity, captured by ENN, ranks third. Equity rises markedly while mean inter-park distance remains below roughly 450 m, a figure that coincides with the walkable-neighborhood radius. Once ENN surpasses that threshold the justice dividend reverses; connected green corridors, infill plazas and rooftop gardens can restore continuity where contiguous land is scarce. Where gated compounds block direct routes, semi-public corridors threading through private estates—negotiated via incentives or design concessions—should extend the effective reach of the network.

Edge geometry also matters. Higher perimeter-to-area ratios and greater outline complexity enlarge the "social edge," multiplying points of contact between parks and their catchments. Compact yet irregular shapes therefore deserve preference whenever parcels are redesigned or new sites are carved from redevelopment projects.

The practical translation of these principles is illustrated schematically in Fig. 2.11, which contrasts multi-centered metropolitan fabrics with single-centered smaller towns and maps appropriate interventions for each. Implementation

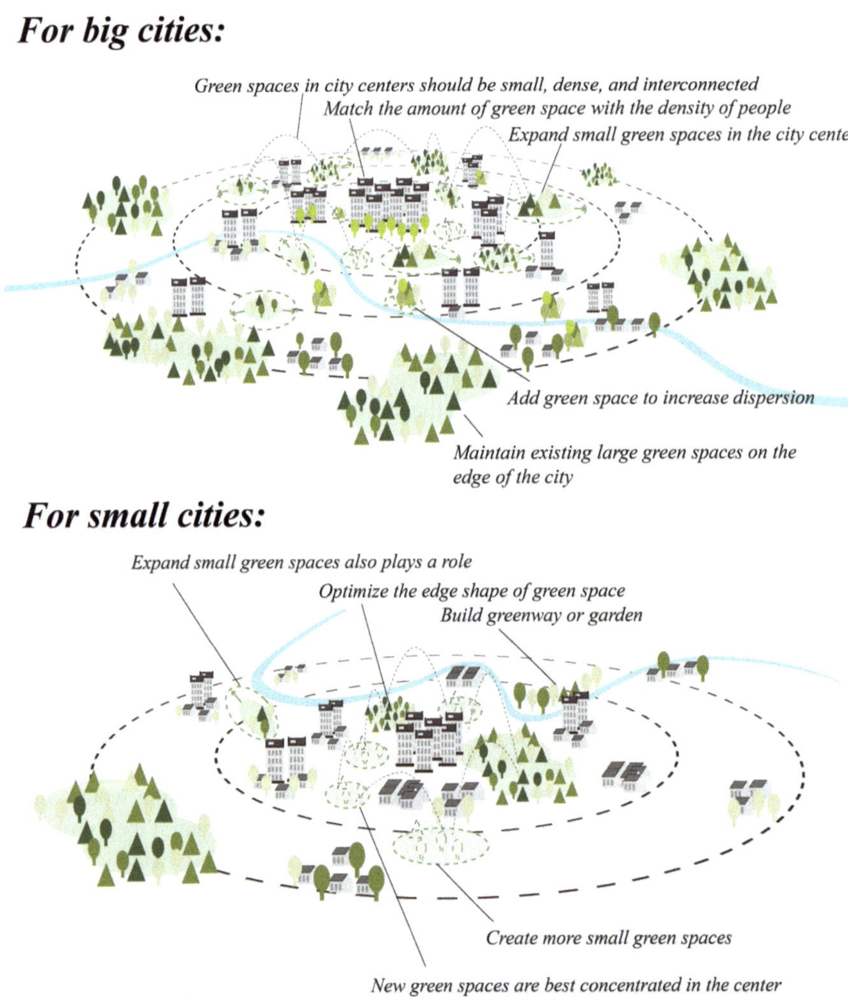

Fig. 2.11 Practical suggestions

nevertheless encounters hurdles: gated compounds restrict permeability; fiscal and administrative resources remain limited in many jurisdictions; and public consensus must still be forged. Participatory planning, already gaining ground in China, is therefore essential. Continuous citizen involvement in design, stewardship, and maintenance strengthens social attachment to parks and guards against quasi-privatized outcomes. Public–private partnerships, phased roll-outs, and targeted subsidies can further alleviate budgetary strain while ensuring that the network of green assets evolves in line with residents' expressed needs and values.

References

Bai X, Chen J, Shi P (2012) Landscape urbanization and economic growth in China: positive feedbacks and sustainability dilemmas. Environ Sci Technol 46(1):132–139. https://doi.org/10.1021/es202329f

Biró TS, Néda Z (2020) Gintropy: Gini Index based generalization of entropy. Entropy 22(8):Article 8. https://doi.org/10.3390/e22080879

Bolitzer B, Netusil NR (2000) The impact of open spaces on property values in Portland, Oregon. J Environ Manag 59(3):185–193. https://doi.org/10.1006/jema.2000.0351

Chen WY, Wang DT (2013) Economic development and natural amenity: An econometric analysis of urban green spaces in China. Urban For Urban Green 12(4):435–442. https://doi.org/10.1016/j.ufug.2013.08.004

Chen Y, Yue W (2020) Which communities have better accessibility to green space? An investigation into environmental inequality using big data. Landsc Urban Plan 204:103919. https://doi.org/10.1016/j.landurbplan.2020.103919

Chen Y, Chen Z, Xu G, Tian Z (2016) Built-up land efficiency in urban China: Insights from the General Land Use Plan (2006–2020). Habitat Int 51:31–38. https://doi.org/10.1016/j.habitatint.2015.10.014

Chen Y, Xu Z, Byrne J, Xu T, Wang S, Wu J (2021) Can smaller parks limit green gentrification? Insights from Hangzhou, China. Urban For Urban Green 59:127009. https://doi.org/10.1016/j.ufug.2021.127009

Choumert J, Salanié J (2008) Provision of urban green spaces: Some insights from economics. Landsc Res 33(3):331–345. https://doi.org/10.1080/01426390802045996

Comber A, Brunsdon C, Green E (2008) Using a GIS-based network analysis to determine urban greenspace accessibility for different ethnic and religious groups. Landsc Urban Plan 86(1):103–114. https://doi.org/10.1016/j.landurbplan.2008.01.002

Fan P, Xu L, Yue W, Chen J (2017) Accessibility of public urban green space in an urban periphery: The case of Shanghai. Landsc Urban Plan 165:177–192. https://doi.org/10.1016/j.landurbplan.2016.11.007

Fleischer F (2007) "To choose a house means to choose a lifestyle." The consumption of housing and class-structuration in urban China. City Soc 19(2):287–311. https://doi.org/10.1525/city.2007.19.2.287

Friedman JH (2001) Greedy function approximation: A gradient boosting machine. Ann Stat 29(5):1189–1232

Fung T, Siu W-L (2001) A study of green space and its changes in Hong Kong using NDVI. Geogr Environ Model 5(2):111–122. https://doi.org/10.1080/13615930120086032

Gu X, Wu Z, Liu X, Qiao R, Jiang Q (2024) Exploring the nonlinear interplay between urban morphology and nighttime thermal environment. Sustain Cities Soc 101:105176

Hastie T, Friedman J, Tibshirani R (2001) The elements of statistical learning. Springer, New York. https://doi.org/10.1007/978-0-387-21606-5

Ke G, Meng Q, Finley T, Wang T, Chen W, Ma W, Ye Q, Liu T-Y (2017) LightGBM: A highly efficient gradient boosting decision tree. Adv Neural Inf Proces Syst 30:3146–3154

Kim Y, Kim Y (2022) Explainable heat-related mortality with random forest and SHapley Additive exPlanations (SHAP) models. Sustain Cities Soc 79:103677. https://doi.org/10.1016/j.scs.2022.103677

Koprowska K, Łaszkiewicz E, Kronenberg J (2020) Is urban sprawl linked to green space availability? Ecol Indic 108:105723. https://doi.org/10.1016/j.ecolind.2019.105723

Kronenberg J, Haase A, Łaszkiewicz E, Antal A, Baravikova A, Biernacka M, Dushkova D, Filčak R, Haase D, Ignatieva M, Khmara Y, Niţă MR, Onose DA (2020) Environmental justice in the context of urban green space availability, accessibility, and attractiveness in postsocialist cities. Cities 106:102862. https://doi.org/10.1016/j.cities.2020.102862

Li X, Zhang C, Li W, Ricard R, Meng Q, Zhang W (2015) Assessing street-level urban greenery using Google Street View and a modified green view index. Urban For Urban Green 14(3):675–685. https://doi.org/10.1016/j.ufug.2015.06.006

Lichtenberg E, Ding C (2009) Local officials as land developers: Urban spatial expansion in China. J Urban Econ 66(1):57–64. https://doi.org/10.1016/j.jue.2009.03.002

Lu Y, Chen R, Chen B, Wu J (2024) Inclusive green environment for all? An investigation of spatial access equity of urban green space and associated socioeconomic drivers in China. Landsc Urban Plan 241:104926. https://doi.org/10.1016/j.landurbplan.2023.104926

Nesbitt L, Meitner MJ, Girling C, Sheppard SRJ, Lu Y (2019) Who has access to urban vegetation? A spatial analysis of distributional green equity in 10 US cities. Landsc Urban Plan 181:51–79. https://doi.org/10.1016/j.landurbplan.2018.08.007

O'Donnell O, O'Neill S, Van Ourti T, Walsh B (2016) Conindex: estimation of concentration indices. Stata J 16(1):112–138. https://doi.org/10.1177/1536867X1601600112

Panduro TE, Veie KL (2013) Classification and valuation of urban green spaces—A hedonic house price valuation. Landsc Urban Plan 120:119–128. https://doi.org/10.1016/j.landurbplan.2013.08.009

Qiao L, Zhuang J, Zhang X, Su Y, Xia Y (2021) Assessing emotional responses to the spatial quality of urban green spaces through self-report and face recognition measures. Int J Environ Res Public Health 18(16):8526

Rigolon A (2016) A complex landscape of inequity in access to urban parks: A literature review. Landsc Urban Plan 153:160–169. https://doi.org/10.1016/j.landurbplan.2016.05.017

Shi X, Cheng Y, Xue D (2019) Classification algorithm of urban point cloud data based on LightGBM. IOP Conf Ser Mater Sci Eng 631:052041. https://doi.org/10.1088/1757-899X/631/5/052041

Wang S, Yang Z, Liu H (2011) Impact of urban economic openness on real estate prices: Evidence from thirty-five cities in China. China Econ Rev 22(1):42–54. https://doi.org/10.1016/j.chieco.2010.08.007

Wang B, Xu T, Gao H, Ta N, Chai Y, Wu J (2021) Can daily mobility alleviate green inequality from living and working environments? Landsc Urban Plan 214:104179. https://doi.org/10.1016/j.landurbplan.2021.104179

Wang B, He S, Ma W (2022) Does park size affect green gentrification? Insights from Chongqing, China. Sustainability 14(16):Article 16. https://doi.org/10.3390/su14169916

Wu L, Kim SK (2021) Does socioeconomic development lead to more equal distribution of green space? Evidence from Chinese cities. Sci Total Environ 757:143780. https://doi.org/10.1016/j.scitotenv.2020.143780

Xiao Y, Lu Y, Guo Y, Yuan Y (2017) Estimating the willingness to pay for green space services in Shanghai: Implications for social equity in urban China. Urban For Urban Green 26:95–103. https://doi.org/10.1016/j.ufug.2017.06.007

Yew CP (2012) Pseudo-urbanization? Competitive government behavior and urban sprawl in China. J Contemp China 21(74):281–298. https://doi.org/10.1080/10670564.2012.635931

Zhu P, Zhang Y (2008) Demand for urban forests in United States cities. Landsc Urban Plan 84(3):293–300. https://doi.org/10.1016/j.landurbplan.2007.09.005

Chapter 3
Research Perspective on City-Scale Urban Green Space and Equity

3.1 Urban Green Space Equity and Its Impacts on Open Communities

Over the last two decades, gated communities, which are walled residential neighborhoods, have become a common feature of metropolitan areas worldwide (Atkinson and Blandy 2005; Lang and Danielsen 1997). Moreover, gated communities have been empirically demonstrated by scholars to reduce public spaces, thereby exacerbating social inequities—a phenomenon particularly detrimental to socioeconomically disadvantaged groups (Webster et al. 2002). In China, such communities remain prevalent but are projected to be progressively replaced by open communities (Gieryn 2000). The 2016 Guidelines on Strengthening Urban Planning and Development Management issued by the Central Committee of the Communist Party of China and the State Council explicitly mandates that new residential developments adopt the block-based system, with a principal prohibition on new gated communities. It further requires existing gated residential compounds and work-unit compounds to be gradually opened. Would dismantling the barriers of gated communities alleviate green space inequity? This section addresses this question through empirical case analysis.

1. Metrics for Community Green Space Accessibility at the Urban Scale
Beijing's central urban area was selected as the case study region to explore whether open communities alleviate green space inequity. Given that residents within the study area rarely travel beyond catchment thresholds to visit urban parks, a Gaussian-based 2SFCA method was employed to assess green space accessibility, continuously differentiating spatial accessibility levels within defined catchment thresholds.

First, we utilized a 10-km buffer zone (Fig. 3.1) that included 947 green spaces in Beijing to address the "edge effect," which is always realized by compressing the accessibility of those who live along the research boundary. In other words, residents near the boundary may stroll about urban green spaces outside the study area.

J. Wu, *Urban Green Space Equity in China*, Urban Sustainability,
https://doi.org/10.1007/978-981-95-5771-4_3

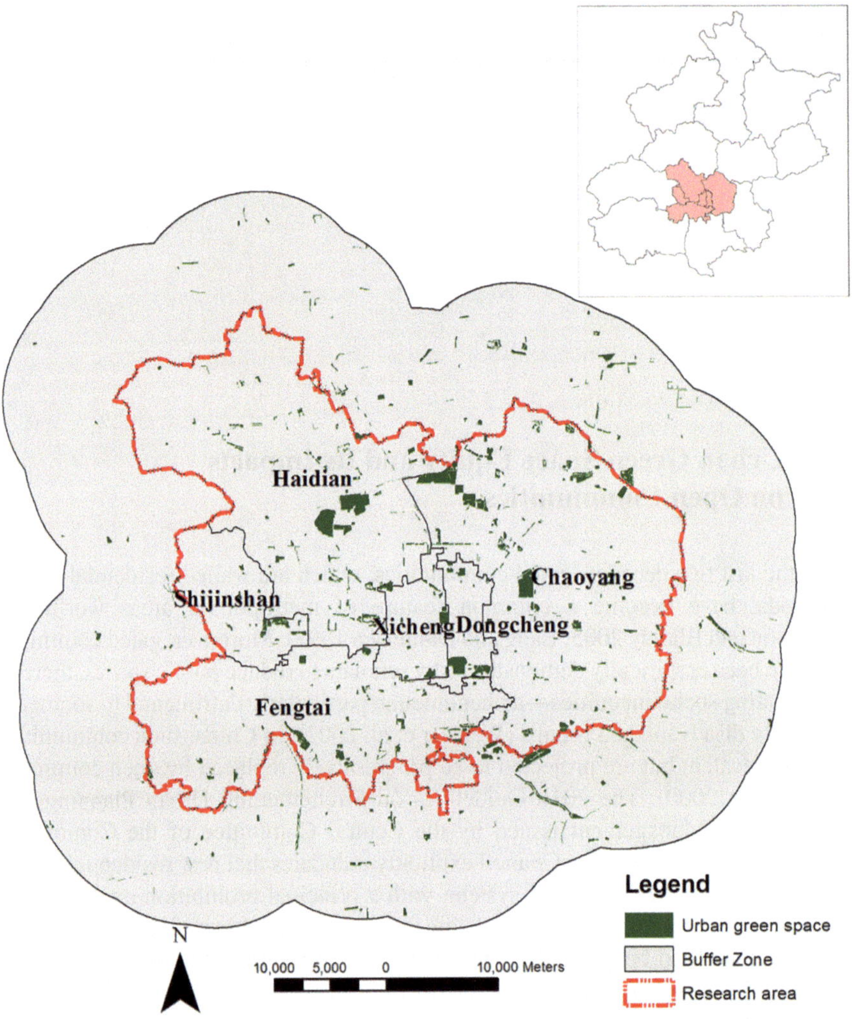

Fig. 3.1 Community and green space sampling scope at urban scale

Residential zone data and their attributes were obtained from the SouFun Holdings website using crawler technology (containing 2961 residential zones, including 83 open residential zones and 2878 residential zones with gates) in our research area. We obtained physical attributes and locations, such as the floor-area ratio of the residential zone, the total households, the greening rate, and the median housing price, of the 2878 observations at the SouFun website (Fig. 3.2). We divide the 2878 residential zones into five groups according to housing price around the following quantiles: >70,000 yuan/m^2, 50,000–70,000 yuan/m^2, 45,000–50,000 yuan/m^2, 40,000–45,000 yuan/m^2, and <40,000 yuan/m^2. Each group contains

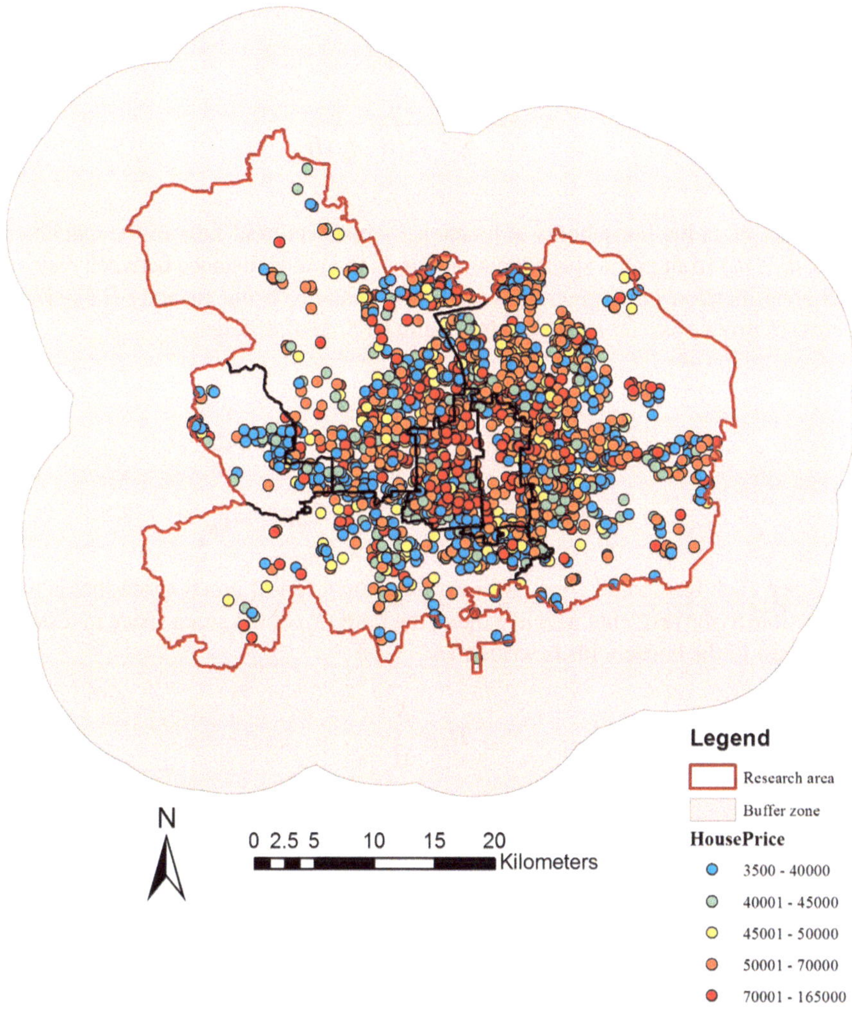

Fig. 3.2 Community locations and average housing prices in urban Beijing

approximately 575 samples. Those living in residential zones with an average housing price below 40,000 yuan were considered low-income groups.

Subsequently, we use the 2SFCA method to calculate accessibility of urban green spaces in Beijing (Dai 2010).

As the first step, for each green space location i, search all residential zone locations (j) that are within a threshold travel distance d_0 from i, thus formulating the catchment for green space location i. In the Beijing urban study, catchment distance thresholds were established for different transportation modes: Walking: 1 km; Cycling: 2.5 km; Driving: 5 and 10 km. The populations at j will be weighted using a Gaussian function (G). Sum up the weighted households within the catchment for

i as the potential users for green space at j. The ratio (R_i) of the green space as parks to households is written as follows:

$$R_i = \frac{S_i}{\sum_{k \in \{d_{ji} \le d_0\}} G(d_{ij}, d_0) H_j}$$

where H_j is the households at location j whose centroid falls in the catchment (i.e., $d_{kj} \le d_0$) from green space location i, d_{ij} is the travel distance between residential zone location j and green space location i, and S_i is the capacity (i.e., size in acres) of green space at i. G is the following friction of distance:

$$G(d_{ij}, d_0) = \begin{cases} \dfrac{e^{-\left(\frac{1}{2}\right)\times\left(\frac{d_{ij}}{d_0}\right)} - e^{-\left(\frac{1}{2}\right)}}{1 - e^{-\left(\frac{1}{2}\right)}}, & \text{if } d_{ij} \le d_0 \\ 0, & \text{if } d_{ij} > d_0 \end{cases}$$

Since each residential zone has a certain proportion of green space that is only accessible to the residents who live there, the ratio (R_j) of the green space in residential zones to the households is written as:

$$R_j = \frac{S_j}{H_j}$$

As the second step, for each residential zone location j, search all parks l within the threshold distance d_0 from j, thus formulating the catchment for the household at j. Discount each R_j using the Gaussian function (G). Sum up the discounted R_j within the catchment area j to obtain the spatial accessibility at residential location j and that household's R_j with no discount as follows:

$$A_j = \sum_{l \in \{d_{jl} \le d_0\}} G(d_{jl}, d_0) R_l + R_j$$

where l denotes all parks within the catchment of residential zone location j, and all other notations are the same as in the first equation. The accessibility score (A'_j) suggests the amount of green spaces (in acres) for every household in a neighborhood.

As the final step, we will calculate A'_j by simulating a situation in which all communities become open by dismantling the fences. Each green space in every residential zone can be accessed by all citizens. Then, the accessibility score (A'_j) is as follows:

$$A'_j = \sum_{l \in \{d_{jl} \le d_0\}} G(d_{jl}, d_0) R_l + \sum_{l \in \{d_{jk} \le d_0\}} G(d_{jk}, d_0) R_j$$

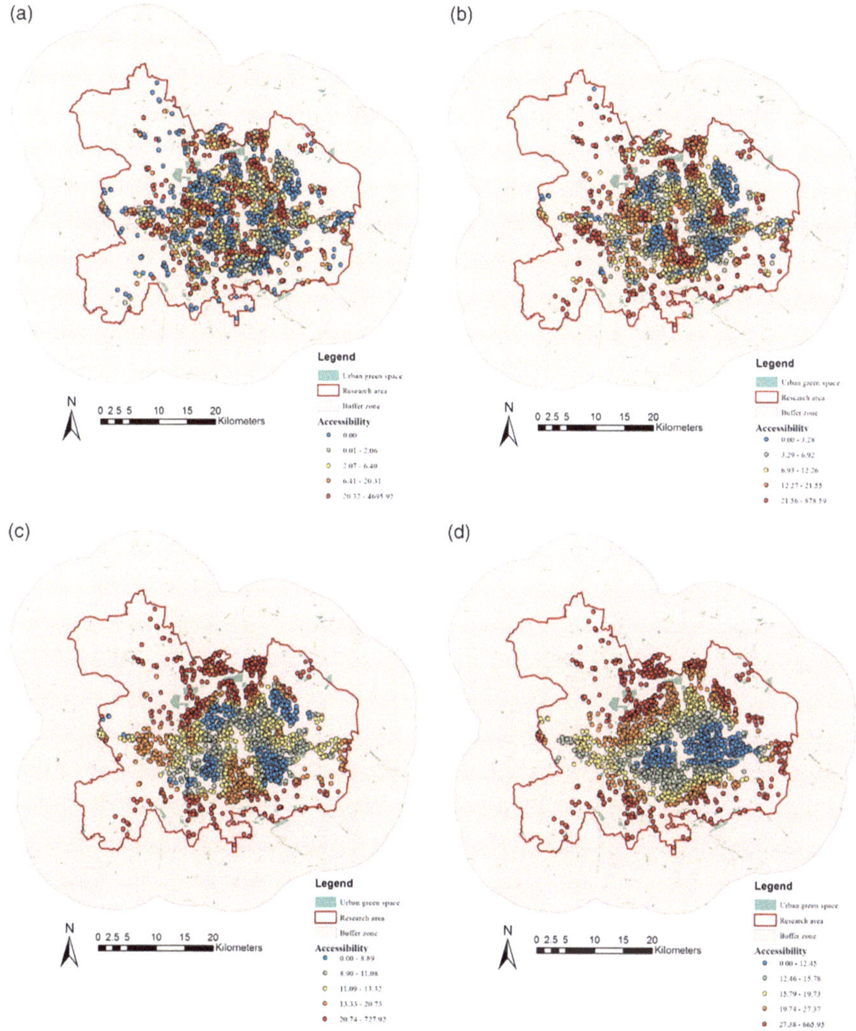

Fig. 3.3 Accessibility to urban green spaces at varying distance thresholds in gated communities of Beijing

where k denotes all attached green spaces in residential zones within the catchment of residential zone location j, and all other notations are the same as in the first equation.

2. Urban-Scale Disparities in Community Green Space Accessibility

1. Accessibility to community green spaces under varying distance thresholds was compared before and after community opening. The results reveal that when communities were gated, green space accessibility was universally prevalent across the study area regardless of distance thresholds (Fig. 3.3). At the 1-km threshold, more accessible green spaces were predominantly concentrated

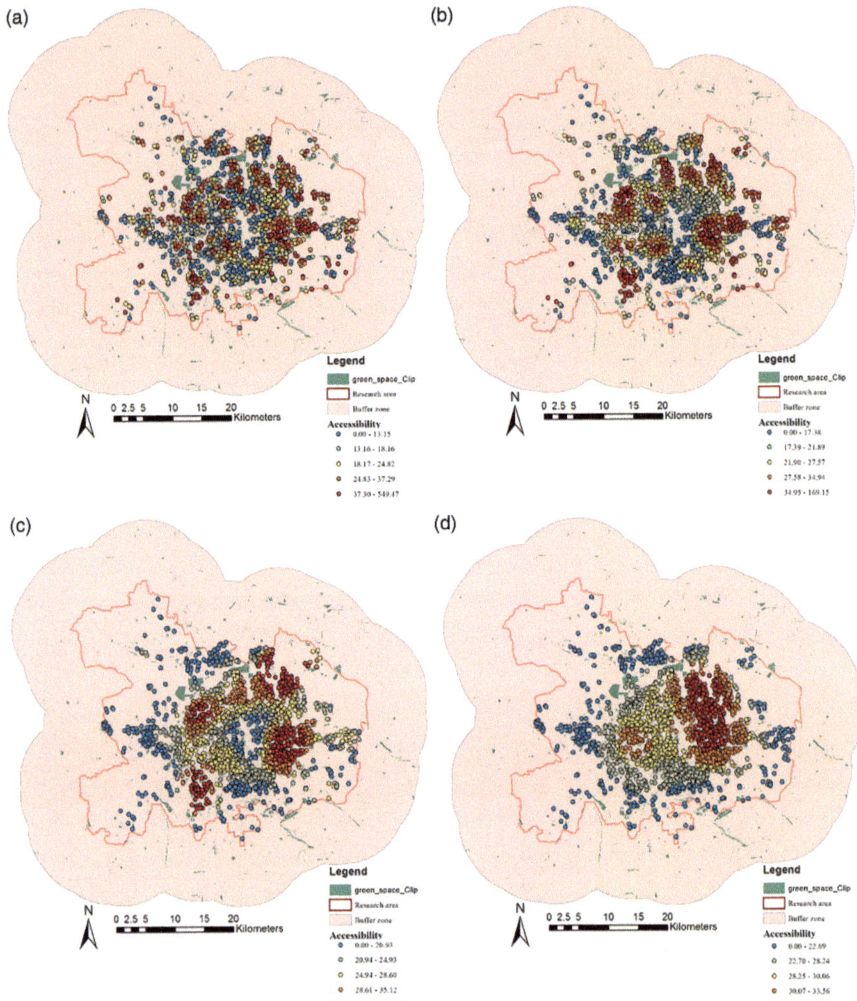

Fig. 3.4 Accessibility to urban green spaces at varying distance thresholds in open communities of Beijing

around major parks (Fig. 3.3). Similarly, the 2.5-km threshold confirmed higher accessibility near parks, yet simultaneously exposed inaccessible zones with minimal or no park access, where even parks-equipped communities suffered from poor transportation connectivity. When thresholds expanded to 5- and 10-km catchment areas, downtown areas exhibited significantly fewer opportunities to access green spaces compared with suburban areas.

We then develop a scenario in which all communities are open to the public. Overall, urban green access increased from a gated state to an open state for each catchment because residential parks were considered in the measurements (Fig. 3.4).

Despite this overall increase, different areas experienced substantially different outcomes. Many residential zones that are scattered in the research area are identified in terms of changes in urban green space accessibility when $d_0 = 1$ km, whereas they are more concentrated when the catchment is 2.5 and 5 km. The pattern presents obvious concentric circles when $d_0 = 10$ km.

2. Multivariate regression models, including the ordinary least squares model, were then employed to evaluate the disparities in accessibility. We set the low-income residential zones (average housing price <40,000 yuan/m^2) as the reference group and the four dummy variables representing the other four groups as the comparison groups. The *p*-value of each coefficient of a dummy variable would be significant if that experimental group differed from the reference group. The results of the regression modeling to assess the patterning of green space availability by residents' socioeconomic circumstances indicate that the green space accessibly of residential zones in Beijing was substantively different depending on the income bracket within the context of gated communities.

The most expensive gated residential zones have significantly more accessible green space than do those whose residents earn the lowest income when the 1-km threshold is applied. However, the access of residents in the second income level is significantly different from that of the low-income groups. Under application of the 2.5- and 5-km thresholds, the same situation appears in which disparities of green space accessibility will develop more significantly when residential zones are open. When the threshold is 10 km, the green space accessibility clearly varies across each income level whether the residential zones are gated or open. These results show that the inequity of access to green space is exacerbated when gated communities are opened up.

3. Underlying Mechanisms Which Community Openness Exacerbates Green Space Inequities

To explore the reasons behind this result, which was the opposite of what we expected, we assessed the spatial pattern of residents' incomes using Anselin Local Moran's I in ArcGIS. In Fig. 3.5, we found a spatial polarization between the rich and the poor. The high–high clusters and low–low clusters are far greater than the high–low and low–high clusters. This finding shows that residential zones near affluent dwelling districts are likely to be sold at very high prices. In other words, affluent residential zones are agglomerated, whereas impoverished residential zones are built together. Although the green space accessibility of low-end residential areas will be enhanced after communities open their gates, it will increase much more in rich residential zones. Therefore, there will be more serious inequities in green space accessibility.

Statistical analysis confirms pronounced inequity in urban green space access among low-income groups in Beijing. Post-community opening, disparities between socioeconomic groups have intensified. Future regulatory detailed planning should stipulate minimum and maximum floor-area ratios (FAR) to restrict developers from constructing high-end residential zones near urban green spaces or prioritize residential land parcels with high green space accessibility for affordable housing developers. These policies would enhance low-income residents' proximity to urban green infrastructure. Concurrently, green space accessibility must be increased

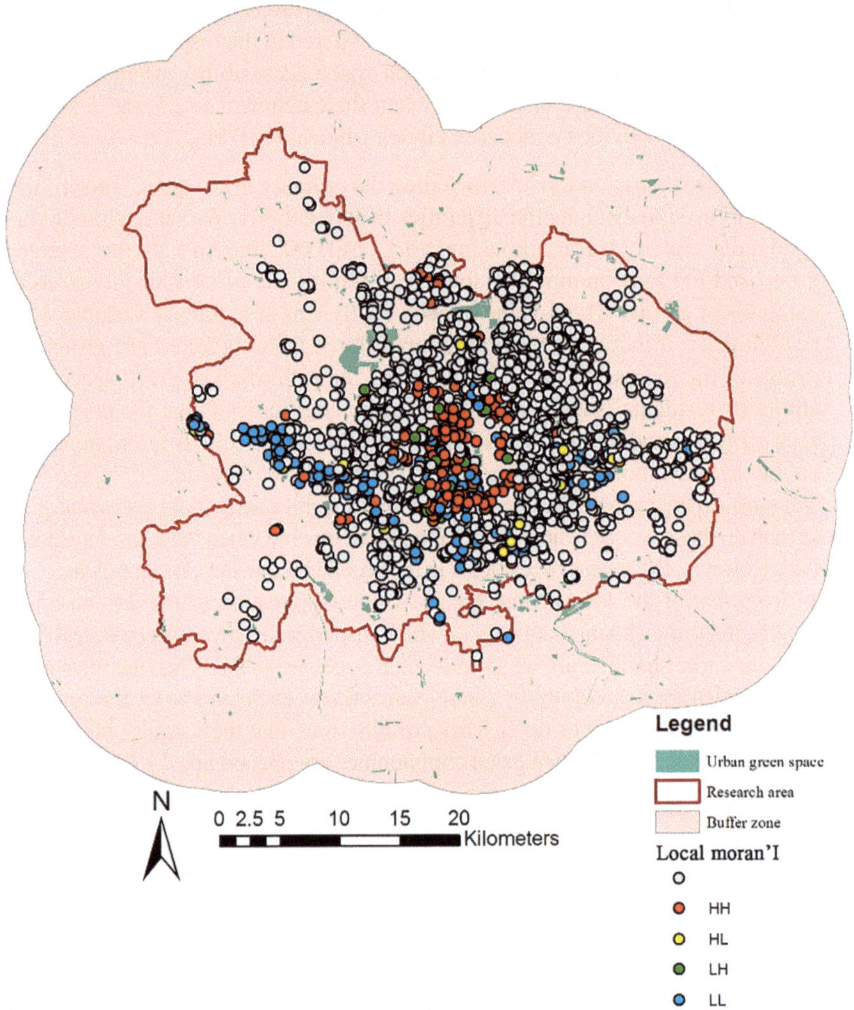

Fig. 3.5 Spatial pattern of resident income in urban Beijing based on cluster and outlier analysis

in areas with higher concentrations of vulnerable groups, directing policy attention to socioeconomically disadvantaged neighborhoods. This methodology advances the evaluation of green space accessibility and equity in Chinese cities, while providing evidence-based recommendations to optimize open community policies.

3.2 Streetscape Greenery Evaluation in Urban Green Space Equity Frameworks

Previous studies on green space equity optimization revealed an overreliance on patchy green spaces (e.g., parks) as proxies for urban greenery, while linear green infrastructure such as street-level greenery was largely overlooked. In reality, linear green elements critically shape human-scale green exposure and perception, as pedestrians inherently interact with streetscape vegetation during daily mobility to green destinations. Consequently, integrating street-level greenery into green equity assessments is essential for developing a comprehensive urban green equity metric system.

1. Measuring Urban Green Space Accessibility Incorporating Street-Level Green View Metrics

Incorporating linear urban green infrastructure into accessibility calculations necessitates a specialized modeling approach (Lee and Hong 2013; Piovani et al. 2018; Rigolon 2016). Conventional methods—including container, coverage, gravity, and 2SFCA models—fail to quantify street-level green accessibility or capture the attractiveness of integrated green networks (Luo and Whippo 2012). This section therefore employs CFCA model to measure accessibility across both areal and linear green spaces, followed by Ordinary Least Squares regression to quantify green equity disparities among socioeconomic residential clusters (Wu et al. 2019).

Following the established methodology, an empirical study was conducted in urban Beijing. Data collection initiated with areal green spaces, where 947 urban green patches served as the basis for generating 10-km buffer zones. Conventional approaches to evaluating street-level greenery—relying solely on field surveys or remote sensing-derived canopy cover images—exhibit critical limitations: on the one hand, prohibitive costs for fine-grained streetscape characterization; on the other hand, significant discrepancies between objective vegetation indices (e.g., NDVI) and human perception of greenness. To address this, Baidu Street View imagery was leveraged with deep learning-based semantic segmentation, capturing human-centric streetscape green perception at scale.

Community-scale data were sourced from 2878 gated communities within Beijing's urban core, previously obtained through web crawling. Communities were stratified into five housing price tiers, with residents in zones averaging <40,000 yuan/m^2 classified as low-income groups for equity analysis.

Following data collection and processing of areal and linear green infrastructure, alongside community datasets in Beijing, we conducted green space accessibility assessment. The conventional 2SFCA model was further refined to propose CFCA model for quantifying accessibility.

The attractiveness of a park is composed of two aspects: its size and the green view index around the park. We first measure the attraction coefficient influenced by size, and we normalize its coefficient:

$$G_{Sj} = \frac{S_j}{\sum_j S_j}$$

$$\overline{G_{Sj}} = \frac{G_{Sj} - \min_{j \in J} G_{Sj}}{\max_{j \in J} G_{Sj} - \min_{j \in J} G_{Sj}}$$

where S_j is the area of urban green space j, G_{Sj} is the size attraction coefficient, and the normalized size attraction coefficient $\overline{G_{Sj}}$ redistributes all the attraction values between 0 and 1.

Based on the size attraction value, each urban green space is assigned an attraction catchment area of a certain size depending on the travel mode. The larger the park is, the larger the extent of its catchment. The catchment area of each park, G_{Sj} is defined using the normalized size attraction coefficient and is dependent on the travel mode that is used. In this study, we will compare spatial accessibility using three modes of transport, (1) driving, (2) Biking, and (3) walking, and translate each normalized size attraction coefficient into a catchment area expressed in kilometers (Table 3.1).

Subsequently, combining the scale-based attraction with the street greenery landscape index yields the comprehensive attraction index for each green space j:

$$G_j = \left[\gamma_s \frac{S_j}{\sum_j S_j}\right] + \left[\gamma_v \frac{V_j}{\sum_j V_j}\right]$$

where the weights γ_s and γ reflect the importance of size (S_j) and street green visual (V_j), respectively, and where $\gamma_s + \gamma_v$ equals 1. Therefore, G_j is the attraction index of each green space j.

Then, we calculated the total demand, D_j for each green space j is calculated by summing the respective populations:

$$D_j = \sum_{i \in \{d_{ij} \le d_0\}} r_i G(d_{ij}, d_0)$$

Table 3.1 The travel mode coefficient translates each normalized attraction coefficient into a catchment area expressed in distance

G_{Sj}	Time(min)	d_0 _Car (km)	d_0 _Biking (km)	d_0_Walking (km)
<0.2	5	2.5	1.25	0.5
0.2–04	10	5	2.5	1
0.4–0.6	20	10	5	2
0.6–0.8	30	15	7.5	3
>0.8	50	25	12.5	5

where r_i is the population in residential zone i, which falls into catchment ($d_{ij} \leq d_0$) for green space j; d_{ij} is the travel distance between residential zone i and green space j, where d_0 is the threshold travel distance; and G is the friction-of-distance listed below:

$$G(d_{ij}, d_0) = \begin{cases} \dfrac{e^{-\left(\frac{1}{2}\right)\times\left(\frac{d_{ij}}{d_0}\right)} - e^{-\left(\frac{1}{2}\right)}}{1 - e^{-\left(\frac{1}{2}\right)}}, & \text{if } d_{ij} \leq d_0 \\ 0, & \text{if } d_{ij} > d_0 \end{cases}$$

We use P_j to present the green-to-population ratio, which is then computed by dividing the attraction value by the total population within the catchment area:

$$P_j = \frac{G_j}{\sum_j D_j}$$

When estimating the spatial accessibility, the park-to-population ratios, R_j, are summed and weighted by the distance that separates the residential zones from the urban green space. A distance-decay function $G(d_{ij}, d_0)$ gives a higher weight to the park-to-population ratio of an urban green space. The spatial accessibility at geographic unit i is then defined as:

$$A_i = \sum_{i \in \{d_{ij} \leq d_0\}} P_j G(d_{ij}, d_0)$$

The spatial accessibility score of each residential zone is estimated three times, once for each mode of transport. The higher the accessibility score, the greater the accessibility of that residential zone to urban green space compared to all other residential zones in the study area.

2. Characterizing Urban Green Space Accessibility with Street-Level Green View Metrics

Figure 3.6 shows the results of the two scenarios and the accessibility scores represented by the different transportation modes. Scenario 1 only accepts urban green space size in the CFCA model as the factor determining the attraction and catchment area. Scenario 2 gives equal weight to the size of urban green space and the street green view around green space.

In the first scenario, accessibility results for walking show that higher accessibility has a dispersed distribution in the research area. Generally, the residential zones near urban green spaces, especially large public parks, received higher accessibility scores. It is surprising that the accessibility results for biking also show significant concentrations. The residential zones in inner cities have remarkably higher

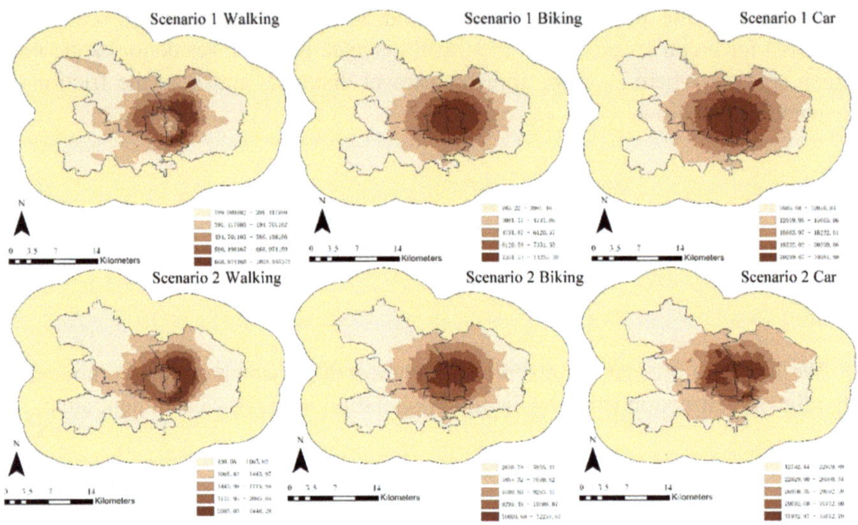

Fig. 3.6 Green space accessibility levels under different transport modes for Beijing communities in dual scenarios

accessibility than those located in suburbs. The same result occurs—and is even more obvious—in the measurement for car mode.

In the second scenario, the agglomeration situation is more obvious but not centralized. Accessibility results show that the residential zones located in the eastern inner city have the highest street greening of the entire research area. In the biking and car transportation model, high accessibility converges in the inner city. There are two reasons for these results. First, the roads in the inner city are old and narrow, while the roads in the suburbs are new and wide, although the rates of green are the same. The street green view is higher on narrow roads than on broad roads. Second, the streets in inner cities are generally greener due to lush roadside trees, while the streets in the suburbs are barren, with only a few young trees.

We employ multivariate regression with the ordinary least squares model. Three transportation modes are adopted to evaluate the disparities in access to urban green space. Low-income communities are set as the reference group and are defined as residential zones with average housing prices lower than 40,000 yuan/m^2. In addition, four comparison groups are set as dummy variables to compare access to green space among different socioeconomic residential zones. If comparison groups and reference groups are significantly different, each dummy variable's coefficient could pass the confidence test. The regression results show that residential zones with different socioeconomic circumstances have different green space availability. For the first scenario, the green space accessibility of residential zones in Beijing was substantially different depending on the income bracket. Whichever transportation mode was chosen, the expensive residential zones obtain higher green accessibility scores than do low-income residential zones. For the second scenario, the results are very different. When residents walk from the most expensive (more than

70,000 yuan/m^2) residential zones, they can access more green space than residents of low-income residential zones. Beyond that, the experiential groups have no better access to urban green space. The situation is similar when the residents choose to bike. If only cars are employed as transportation, the disparities in accessibility for urban green space are salient based on economic level.

By integrating street-level linear greenery into urban green space characterization, we optimized the mathematical framework for accessibility measurement. This revealed a counterintuitive finding: incorporating street greenery significantly mitigates green inequity manifestations. This unlocks a novel pathway for addressing urban green injustice—enhancing street-level green landscapes and upgrading linear green infrastructure quality could potentially recalibrate green service provision toward greater spatial equity.

3.3 Integrated Service Capacity of Green Spaces in Urban Equity Contexts

Integrating linear street greenery into urban green space characterization offers a potential pathway to mitigate green injustice through refined equity assessments. However, this approach remains constrained by oversimplified representations of green space attractiveness—traditionally reduced to mere size and Green View Index (GVI)—while neglecting multifaceted service attributes (Li et al. 2015). Prior studies exhibit analogous limitations: despite continuous methodological advances in accessibility modeling (e.g., 2SFCA variants), the field persists in evaluating quantity-based accessibility, where attractiveness is monolithically quantified via area or canopy volume metrics (Wu et al. 2022). In reality, urban green spaces exhibit heterogeneous features—facility diversity, vegetation configuration, species richness—that critically shape human green experience. Persisting with quantity-centric paradigms not only impedes accurate assessment of quality-based accessibility (reflecting integrated user experiences) but also risks misestimating inequality magnitudes, thereby confining quality-enhancement strategies to qualitative vagueness.

Therefore, further in-depth research on urban green space equity necessitates an optimized characterization of green space's comprehensive service capacity. The fundamental research approach involves: (1) establishing an evaluation system for comprehensive green space service capacity (incorporating greenspace visibility index, the scale of areal green spaces, walkability, connectivity, diversity, ecological service capacity, and compactness); (2) employing the CFCA model proposed in the previous section; and (3) combining the Gini index, spatial cluster analysis, and multinomial logistic regression. This integrated methodology will be used to compare the green space accessibility and equity performance of urban communities in Beijing under three distinct scenarios (Scenario 1: considering only green space scale; Scenario 2: considering green space scale and street greenery; Scenario 3:

considering comprehensive green space service capacity and street greenery) at varying distance thresholds across different transportation modes.

1. Assessment Framework for Green Space Comprehensive Service Capacity
First, a framework was established for assessing green space comprehensive service capacity and calculating accessibility (Fig. 3.7), followed by the acquisition of relevant green space and community data for urban Beijing. Notably, the indicators used to characterize green space quality were carefully selected. To select the index that can truly reflect the value of green space to residents, we first searched the empirical research and review journal papers published in Chinese and English from 1990 to 2020 for works including "green space quality" in abstract or keywords. According to the abstract and content of the article, we removed the papers that have nothing to do with residents' well-being (such as soil fertility, land asset value, etc.) and the research on specific sick groups or elderly groups with limited movement. We summarized the description indicators of green space quality in these papers (which are of positive significance to residents' well-being). We then removed the subjective perception indicators that cannot be measured by the objective environment (such as quiet and relaxation) and the detailed or highly specific green space design features (such as flower bed design and facility design features). For the remaining indicators, we used natural language processing for word frequency analysis and selected the indicators higher than 30% quantile threshold. Considering that time and regional differences may make these indicators unsuitable for the Chinese environment, we invited 30 scholars from the Urban Planning

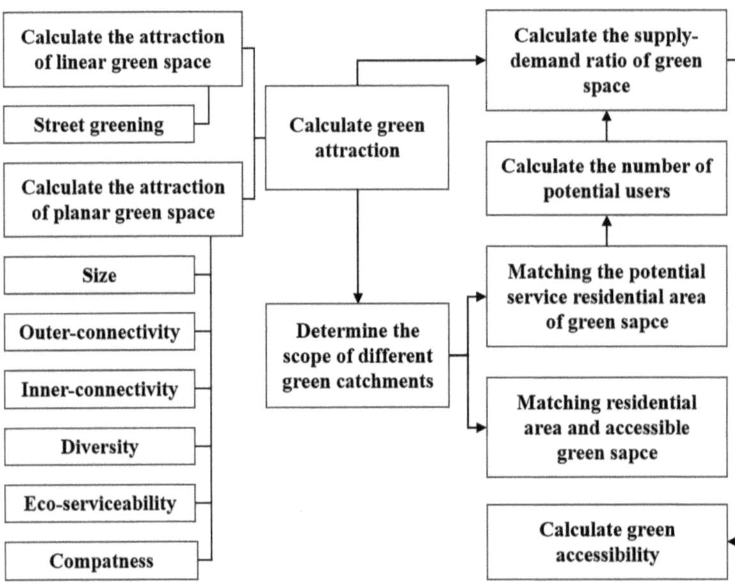

Fig. 3.7 Computational framework for green space accessibility under comprehensive service capacity assessment

Bureau or engaged in urban green space research to judge the operability of the project and classify the indicators, remove the inapplicable items, merge similar definition items, adjust the specific standards of the indicators (such as buffer size, land classification, etc.), and finally summarize them into a multi-level indicator framework reflecting six dimensions.

The specific data collection methods for indicators related to green space service capacity (quality) and community data are described as follows:

1. Urban green space data: The vector data of green space is mainly from the land-use survey (2010). After that, the change of land use within the time difference (2010–2015) is checked according to the land-transfer information published by the Ministry of Natural Resources on the official website. The data of urban green space area are obtained from the Beijing Urban Green Space System Plan (2004–2020), including the green space scale built and planned. For the green space not included in the Plan, it is obtained from the Beijing public data open platform (https://data.beijing.gov.cn/). The platform contains the information of Beijing registered parks, including attribute fields such as serial number, name, district, street, address, and consultation telephone, which are provided by Beijing Landscaping Bureau. A total of 826 green areas of more than 2 ha were obtained in the study area.
2. Residential zone data: The residential zone data are obtained from the SouFun website (https://www.sofang.com/) in 2017 with crawler technology as described in Sects. 3.1 and 3.2. We ultimately obtained the average house price information of 2878 residential zones in our study area, and use it as an agent for income.
3. Road network data: The road network data of walking, cycling, and driving routes from residential zones to each green area were obtained through Amap (https://www.amap.com/) in 2015. And the road network data includes each road grade, length, and location. Amap is the second largest online map supplier in China. Compared with Baidu map, it has better traffic navigation function advantages and has more complete coverage of major cities in China than Google and OpenStreetMap. Considering that most residents go to green space for leisure activities, referring to the opening hours of most parks, the time period for us to obtain the route is concentrated at 8:00–17:00 on weekends. We chose to conduct real-time navigation route simulation when there is little or no congestion on the road and selected the preferred scheme.
4. Street view images data: Baidu Street View images and recognition data identical to those utilized in the preceding section were used. Specifically, green view ratio data derived from machine learning processing, training, and recognition (using the Cityscapes dataset) applied to approximately 500,000 street view images of Beijing were employed.
5. Remote sensing image data: To obtain high-precision remote sensing image data of plant growth period (June to August), we collected data from ESA in 2017 (https://scihub.copernicus.eu/), which was obtained from Sentinel-2 public remote sensing image database of June 2016. We used support vector machine supervised classification in ENVI to interpret the land cover image of remote

sensing image, which is divided into five kinds of land: forest land, grassland, water, construction land, and unused land. We tested the accuracy tested by verifying the sample area of interest and evaluated by confusion matrix. The overall accuracy is 86.22%.

6. POI data: These data are obtained from online map application (https://map.baidu.com/) in 2015.
7. Land-use data: The plot vector graphic data come from the second land survey (2010). According to the official classification standard of land use, it is divided into eight kinds of land: commercial land, industrial and mining storage land, residential land, public management and service land, transportation land, water and water conservancy facilities land, urban villages and industrial land, agricultural land, and others (cultivated land, garden land, forest land, grassland, special land, and other unused land). According to POI data (2015), we took the type proportion of more than 50% as the evaluation standard of unit functional properties and corrected the plots whose properties may change.

Then, by setting the catchment size of different attractive green spaces for different travel modes (walking, cycling, and driving), CFCA model screens the green spaces within the distance threshold of residential areas (determine the distance by calling the route planning function of Baidu map API) and weights and aggregates the green space indicators to obtain the accessibility value. Therefore, the calculation process of different travel modes is the same, and the difference lies in the size of distance threshold.

When evaluating the quality of green space, this study focuses on the ability of green space to provide all aspects of services for residents' leisure activities, involving all stages of residents from transportation to use. Therefore, in order to better reflect the quality of green space recreational serviceability, we collected entries related to the evaluation of green space serviceability based on the existing research literature and selected the entries with a frequency of >30% through word frequency analysis to better reflect the quality of green space. Finally, six representative indicators, namely, size, outer- connectivity, inner-connectivity, diversity, eco-serviceability, and compactness, were selected as the primary indexes to evaluate the quality of green space. Among them, size reflects the tourist carrying capacity and capacity of the green space, outer-connectivity represents the convenience and accessibility of the traffic environment around the green space, inner-connectivity describes the tour line connectivity, diversity reflects the diversity of landscape types, eco-serviceability is used to describe the ecological benefits, and compactness is used to describe the compactness of land use and land construction intensity around the green space. Table 3.2 shows the index framework and its calculation method of green space serviceability.

The first- and second-order mutation index (Table 3.2) are combined for analysis. Based on the theory of fuzzy mathematics, the catastrophe model can greatly reduce the uncertainty and subjectivity of multi-level index evaluation by weighing the relative importance of each index rather than giving weight and index integration (Su et al. 2011, 2018). The specific process of index aggregation is as follows:

Table 3.2 Calculation methodology for comprehensive service capacity indicators of urban green spaces

Primary indicator	Secondary indicator (catastrophe model)	Tertiary indicator	Equation	Explanation	Data source
Recreational serviceability	Size (Cusp)	The total area (TA)	/	The total area of the green park (m^2)	Beijing green space planning (2004–2020) and Beijing registered Park Information (2015).
		The actual usable area of the green park (AUA)	/	The actual usable area of the green park (area of water area removed, m^2)	
	Outer-connectivity (Butterfly)	Width of low-grade Road (LGRW)	/	Average width of tertiary roads and below (weighted by road length), setting 30 m for tertiary roads, 20 m for quaternary roads (residential and unclassified roads), and 10 m for unclassified roads (sidewalks, trails, bicycle lanes)	500 m buffer zone around green space, road data was from Amap
		Road density (RD)	/	Road length per unit area in buffer zone (m/m^2)	
		Intersection density (ID)	/	Number of intersections per unit area in buffer zone (m^{-2})	
		Bus station density (BSD)	/	Number of bus stations per unit area in buffer zone (m^{-2})	500 m buffer zone around green space, road data was from Baidu
	Inner-connectivity (Swallowtail)	Local Integration (LI)	$\mathrm{LI}_i = \dfrac{n\left(\log_2\left(\dfrac{n+2}{3}-1\right)+1\right)}{(n-1)\lvert \mathrm{MD}_i - 1\rvert}$	n is the total number of immediate and non-immediate neighborhood segments that intersect segment i within a given radius; MD is the average depth; i is the segment	50 m buffer zone around green space, calculated in Depthmap road data was from Amap
		Global Integration (GL)	$\mathrm{LI}_i = \dfrac{m\left(\log_2\left(\dfrac{m+2}{3}-1\right)+1\right)}{(m-1)\lvert \mathrm{MD}_i - 1\rvert}$	m is the total number of immediate and non-immediate neighborhood segments that intersect segment i within a given radius; MD is the average depth; i is the segment	
		Depth Distance (MD)	$\mathrm{MD}_i = \dfrac{\sum_{j=1}^{n} d_{ij}}{n-1}$	d_{ij} is the shortest distance between segment i and segment j, n is the total number of the segments	

(continued)

Table 3.2 (continued)

Primary indicator	Secondary indicator (catastrophe model)	Tertiary indicator	Equation	Explanation	Data source
	Diversity (Cusp)	Splitting Index (SPLIT)	$\text{SPLIT} = \frac{A^2}{\sum_{i=1}^{n} a_i^2}$	a_i is the perimeter and area of the patch (class i) in the landscape; A is the total landscape area (m^2)	10 m buffer zone around green space. Remote sensing image from Sentinel-2 (2016). Patches were divided into 5 kinds of land (see supplement for details), calculated in FRAGSTATS
		Shannon's Diversity Index (SHDI)	$\text{SHDI} = -\sum_{i=1}^{m} \left(P_i \ln P_i\right)$	P_i is the type proportion of landscape patch i	
	Eco-serviceability (Butterfly)	Greenspace coverage (GC)	/	Proportion of forest land and grassland area (%)	
		Patch Density (PD)	$\text{PD} = \frac{n_i}{A}(10{,}000)(100)$	n_i is the number of patches (class i) in the landscape; A is the total landscape area (m^2)	
		Euclidean Nearest Neighbor Distance (ENND)	$\text{ENND} = h_{ij}$	h_{ij} is the distance from patch i to the nearest neighboring patch of the same class, based on the edge-to-edge distance	
		Area-Weighted Shaped Index (AWSI)	$\text{AWSI} = \sum_{i=1}^{n} \left[\left(\frac{P_i}{\sqrt{4\pi}\sqrt{a_i}}\right)\left(\frac{a_i}{A}\right)\right]$	P_i and a_i are the perimeter and area of the patch (class i) in the landscape; A is the total landscape area (m^2)	
	Compactness (Swallowtail)	Simpson's diversity index (SDI)	$\text{SDI} = 1 - \sum_{k} \left(\frac{n}{N}\right)^2$	N is the total number of types; n is the number for a particular land use type	Data was from the second land survey (2010) and was revised by POI from Baidu, patches were divided into 8 kinds of land
		Entropy index (EI)	$E = \frac{\sum_{i=1}^{n} P_i \ln\left(P_i\right)}{\ln\left(n\right)}$	n is the total number of land-use types; P_i is the proportion of land use i	
		Possible destination density (PDD)	/	Number of POIs per unit area in buffer zone (m^{-2})	Baidu

Table 3.3 Normalization formula of different catastrophe models from secondary to primary indicators.

A	Control parameter dimensions (*X*)	Normalization formula for response variable (*Y*)
Fold	1	$Y = \sqrt{X_1}$
Cusp	2	$Y = \left(\sqrt{X_1} + \sqrt[3]{X_2}\right)/2$
Swallowtail	3	$Y = \left(\sqrt{X_1} + \sqrt[3]{X_2} + \sqrt[4]{X_3}\right)/3$
Butterfly	4	$Y = \left(\sqrt{X_1} + \sqrt[3]{X_2} + \sqrt[4]{X_3} + \sqrt[5]{X_4}\right)/4$

after the min-max normalization of each index, the variance contribution of each secondary index is compared through factor analysis to determine the importance of the index and bring it into the normalization formula of mutation model according to the order of importance (Table 3.3) so as to complete the aggregation of secondary index to primary index. Where X1 represents the secondary index with the highest importance in factor analysis, and so on. Owing to the obvious collinearity between the six primary indicators (KMO value 0.5699, Bartlett spherical test *p*-value <0.001), they are divided into six principal components (cumulative explained variance ratio 100%) through principal component analysis (PCA), and the variance maximum skew rotation is carried out. The weight is determined according to the load number in the principal component, the characteristic root of the principal component, and the variance contribution rate of each indicator. Following, weighted aggregation is carried out to eliminate the correlation influence of the original variables. Finally, the quality index R_{gj} of green space is obtained.

We used the street green view as the specific evaluation index of street greening around the green space. Street green view refers to the pixel proportion of green vegetation in street view image, which is used to reflect the pedestrian's perception of vegetation during walking. We calculated the average street green view index around the green space based on the road network data containing the green view value of each road (Fig. 3.8). Taking the road length as the weight, we calculated the average green visual rate of the road within the 500 m buffer zone of the green space. For green space *j*, its street greening attraction V_j is as follows:

$$V_j = \frac{\sum_n L_{jn} \cdot \frac{\sum_m V_{jmn}}{m}}{\sum_n L_{jn}}$$

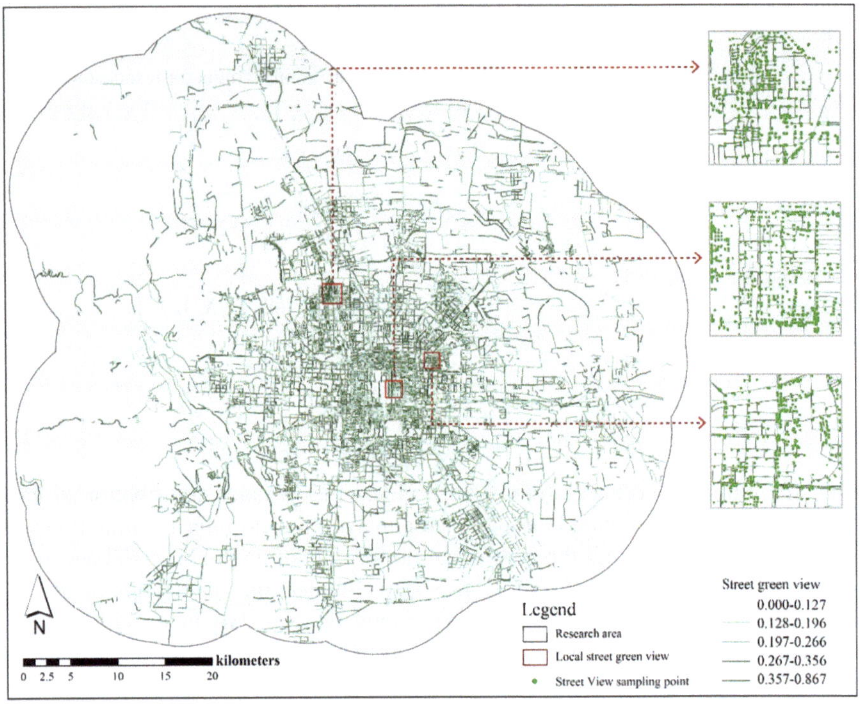

Fig. 3.8 Street green view in urban Beijing

where V_{jmn} represents the green view value of the m-th street view on the n-th road within the catchment around green space j, L_{jn} is the length of the n-th road.

After using the CFCA model to estimate the accessibility features of green spaces, we analyze the disparities in accessibility, which constitutes an assessment of urban green space equity. First, the Gini index is used to evaluate the degree of inequality under two scenarios and three modes of transportation. The accessibility of 2878 communities is ranked from small to large under six scenarios. According to the Gini index formula:

$$\text{Gini} = \sum_{i=1}^{n} H_i A_i + 2\sum_{i=1}^{n} H_i\left(1 - C_i\right) - 1$$

where H_i represents the proportion of households in residence i in all residential households, A_i represents the proportion of the total accessibility of residents in residence i in the total accessibility of residents in all residences, C_i represents the cumulative accessibility ratio ($i = 1, 2, 3, \ldots, n$), and n is the number of residences.

Second, establish Thiesson polygons of green space quantity index and green space quality index. Following, conduct Anselin Local Moran's I analysis on the green space area index, green space serviceability, and green space accessibility under six scenarios to identify the spatial clustering patterns of green space characteristics and residential accessibility in the study area. Finally, according to the

Table 3.4 Indicator contribution to comprehensive service capacity of green spaces

Primary index layer	Secondary index layer	Factor contribution (%)
G_{rj}	Eco-serviceability	16.3
	Size	12.5
	Diversity	12.3
	Inner-connectivity	10.0
	Compactness	35.7
	Outer-connectivity	13.2

average house price, the residential area is divided into five groups like before. Among them, the residence with the average house price lower than 40,000 yuan/m^2 is set as the reference group. Taking the house price as the dependent variable and the green space accessibility value as the independent variable, a multiple logistic regression is constructed to evaluate how the social differentiation caused by income affects the green space accessibility of the residential area.

2. Characterization of Urban-Scale Green Space Comprehensive Service Capacity

1. Green attraction. The contribution of each indicator to the serviceability G_{rj} is obtained through variance decomposition. The results are shown in Table 3.4. The factor contribution rate of size is 12.5%, ranking only the fourth among the six primary indexes.

Figure 3.9 shows the results of Thiesson polygon and spatial cluster analysis of green space area and serviceability. The spatial agglomeration degree of the green space area within the study area is relatively not obvious. By contrast, the serviceability of the green space presents a significant agglomeration, which is manifested in the spatial mode of high in the south and low in the north. It has an obvious overlap with the boundary of the administrative region: low–low clusters are generally distributed in Haidian District, whereas high-high clusters are almost distributed in Fengtai District.

2. Green accessibility. Figure 3.10 shows the spatial distribution and cluster analysis results of green accessibility of two scenarios and three modes of transportation. The result has two main characteristics. On the one hand, quantity accessibility (scenario 1) and quality accessibility (scenario 2) have very different spatial distribution patterns, which are specifically reflected in the sensitivity to transportation modes. Quantity accessibility has obvious spatial agglomeration only in the walking mode, whereas quality accessibility shows significant spatial agglomeration in active travel (walking and cycling). Conversely, regardless of whether scenario 1 or 2, if spatial agglomeration exists, it shows a circular agglomeration distribution with low center and high periphery. In the same study area, even if there are differences in the evaluation system of green space, it can have a more robust spatial clustering pattern.

3. Urban-Scale Equity in Green Space Comprehensive Service Capacity

Calculate the Gini index in six scenarios and obtain the rank from high to low: quantity_walking (0.527160) > quality_walking (0.498422) > quantity_biking

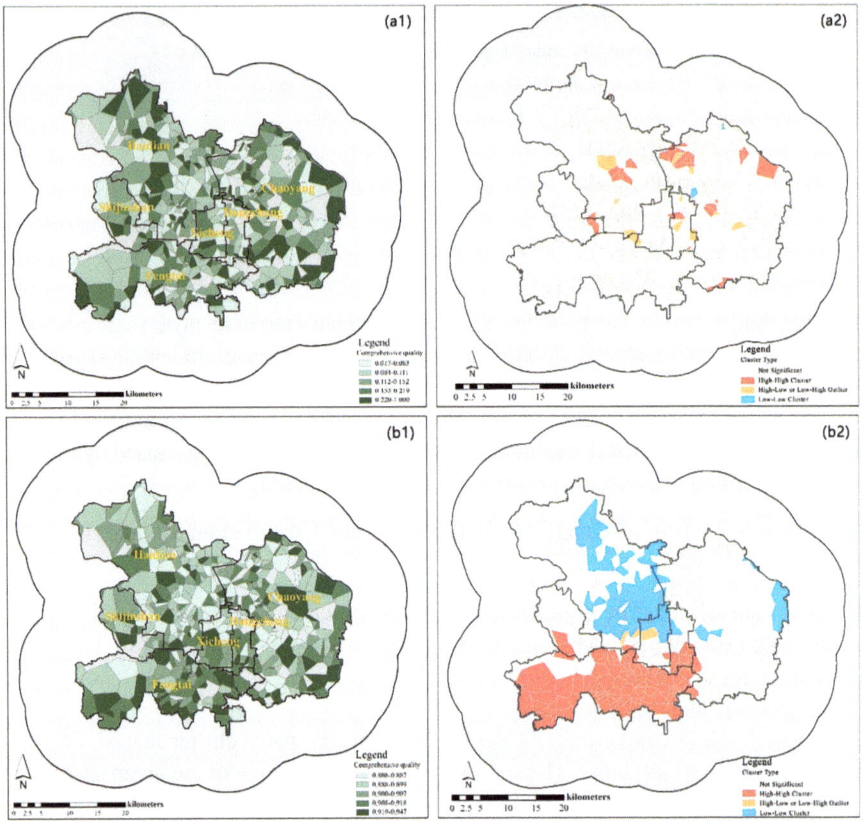

Fig. 3.9 Voronoi polygons and spatial cluster analysis of green space area and service capacity in urban Beijing

(0.452411) > quality_biking (0.397447) > quantity_driving (0.276916) > quality_driving (0.276895). Inequality decreases with the increase of transportation capacity (the expansion of catchments). In addition, quality inequality is always less than quantity inequality, but it is almost consistent in driving mode.

According to the multinomial logistic regression results, when only green space area was considered (i.e., without incorporating street greenery and comprehensive service capacity into the model), high-price residential areas exhibited higher green space accessibility than low-price areas, regardless of the transportation mode selected. Accessibility showed no significant disparity across communities of different price tiers when either walking or cycling was chosen. In contrast, a slight yet statistically significant socioeconomic influence on accessibility emerged under driving mode when evaluating quantity-based accessibility.

The results show that only evaluating the quantity inequality of green space will overestimate the real green space inequality, and the quantity inequality of green space cannot replace the quality inequality. In particular, under the mode of active

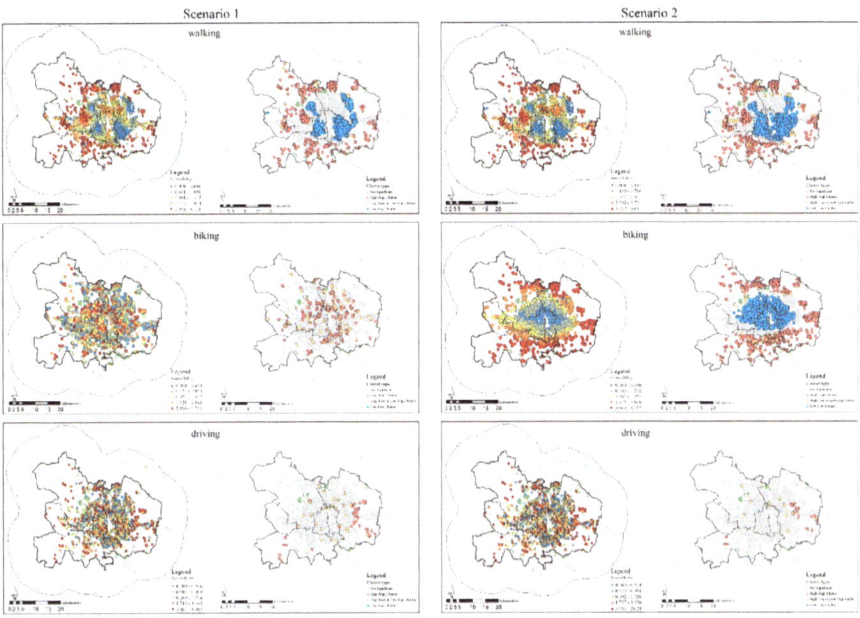

Fig. 3.10 Green space accessibility under different transport modes for Beijing communities in Scenario 1 and 2

travel (walking and cycling), the inequality gap between them is more obvious. As mentioned above, the "real green inequality" in this study emphasizes the inequality of opportunities and benefits for residents to use green space for leisure activities, which is difficult to be reflected only by the size of green space. The neglect of other functions of green space in green inequality assessment hinders people's understanding of the actual problem of green justice and even leads to the measurement deviation of green inequality.

It means that in terms of the spatial pattern of green space attraction, compared with the green space size (scenario 1), the obvious regional aggregation mode of green space quality (scenario 2) means that there is convergence in the green space quality in the same administrative region. After considering the distribution of potential green space users and residential areas, the clustering pattern of green space accessibility has changed significantly from the attraction of green space, and there is pattern robustness within a certain catchment area, especially the low accessibility of the central urban area. This is not only the result of spatial interaction but also the result of the disconnection between the urbanization process at the administrative level and green space development and the management difference between administrative districts. Therefore, government-led green space interventions aimed at enhancing other green space attributes to increase attractiveness and compensate for quantitative deficiencies are likely to narrow disparities in green space service capacity across socioeconomic strata, thereby mitigating inequality arising from socio-spatial differentiation.

In addition, given the inequality caused by the difference of residents' traffic capacity, with the enhancement of traffic capacity, the inequality of green accessibility will weaken and reduce the exclusivity of green space so that more residents have the opportunity to share various types of green space. However, although driving has a greater choice of green space than active travel, as a high proportion of car travel is not conducive to environmental sustainability, both governments and scholars prefer green space to maximize the effect and reduce inequality under active travel mode. On the premise of advocating green travel, enhancing residents' traffic travel capacity and expanding the distance threshold are effective ways to alleviate inequality.

3.4 Optimization Pathways for Advancing Urban Green Space Equity

Following the progressive refinement in characterizing green space attractiveness and empirical validation of green equity performance in Beijing, it becomes imperative to address the demonstrated inequitable supply of green services. A systematic framework must be developed to enhance green equity, thereby offering actionable references and policy recommendations for optimizing urban green governance and advancing social justice in China.

Prompted by these inquiries and aligned with the policy directive of "gradually opening gated residential compounds and institutional campuses" articulated in the 2016 Guidelines on Strengthening Urban Planning and Construction Management issued by the CPC Central Committee and the State Council, this study embarks on an exploration taking Beijing as a case study. The aim is to identify actionable strategies for enhancing green equity through community opening initiatives (Wu et al. 2018; Yu 2017). Literature review reveals a critical gap: scant research exists on quantitative optimization methodologies for green equity. To address this, we propose a dual-method framework: Quantifying accessibility equity dynamics using 2SFCA model to evaluate supply–demand shifts before and after community opening; Generating evidence-based optimization pathways by integrating genetic algorithms to refine spatial intervention priorities.

1. Measuring Accessibility and Equity of Community Green Space at the City Scale

First, we collected the data consisted of the distribution of urban green space and community and road networks in Beijing. Green spaces mainly include urban public and private green spaces in the community. Referring to the criteria of urban green space division in China, we selected urban parks and squares, green spaces with sidewalks along the river/lake, and urban scenic woodlands greater than 2000 m^2. These are the main types of green space for Chinese cities to open to the public and provide ecological, recreational, social, and cultural services. We extracted green patches from the 2009's land survey database, and added new green patches using Baidu Map through manual visual digitizing. Community data was obtained from

Anjuke, China's largest real estate rental service network platform (www.anjuke.com). Web crawler technology was used to capture the community information listed on the website, including the name, location (longitude and latitude), floor area, total number of households, housing price, and greening rate. We calculated the green area of each community according to the floor area × greening rate. We downloaded road network data covering the research area from OpenStreetMap (http://www.openstreetmap.org), a free, open-source, editable global map service provider. The data were checked by topology and connected by road nodes, and the road network was constructed in ArcGIS 10.2 software. The network structure data with nodes and edges were obtained to facilitate network distance analysis.

Second, after collecting the data, a method framework was designed (Fig. 3.11). First, the 2SFCA method will be used to measure the accessibility level of each community in closed and open situations. Next, the Gini index will be introduced to measure the equity of green accessibility. Finally, the GA will be used to optimize the community opening scheme under the goal of promoting equity of green accessibility.

1. Community accessibility metrics. Consistent with the methodology applied in Sect. 3.1, green space accessibility for each community was calculated, along with the degree of change in accessibility before and after community opening initiatives. When communities are opened, each community resident can enjoy free access to other communities' green areas, increasing their options. The change in green accessibility for each community before and after the implementation of the opening policy can be calculated using the following equation:

$$\Delta AF_i = AF_{i\_O} - AF_{i\_C}$$

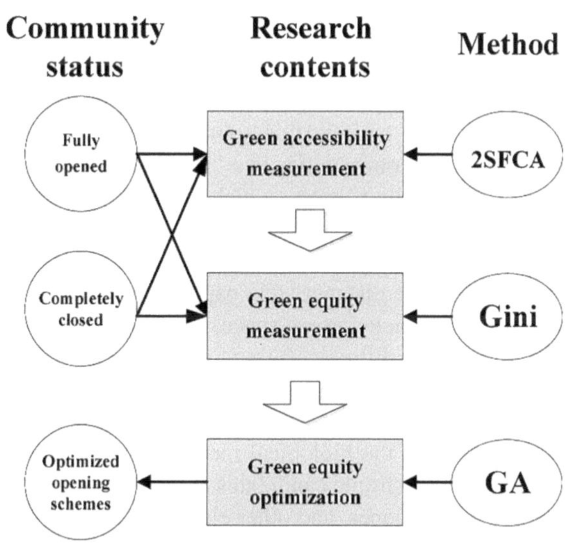

Fig. 3.11 Methodological process framework

In the above equation, AF_{i_O}, AF_{i_C} represent green accessibility of the i-th community after and before being opened, respectively. When the accessibility measurement is completed, we first detect the correlation between green space accessibility and housing price levels by using the Pearson coefficient to test whether green accessibility varies among different income groups.

2. Green space equity dynamics. This section employs the Gini index to measure changes in green equity. By substituting traditional household units with community units and replacing income levels with green accessibility, we adapt the Gini index calculation framework to quantify spatial justice in green resource allocation:

$$GC = 1 + \sum Y_i P_i - 2\sum\left(\sum P_i\right)' Y_i$$

In the above formula, GC represents the Gini index, Y_i represents the proportion of green accessibility in group i out of total green accessibility, and P_i represents the proportion of the number of households in group i out of the total number of households. $(\sum P_i)'$ is the cumulative number of households from Group 1 to Group i as a proportion of the total number of households.

Similarly, the change in green accessibility equity enjoyed by residents before and after the opening of the community can be measured by the following equation:

$$\Delta GC = GC_O - GC_C$$

In the above formula, GC_O and GC_C represent green equity after and before the implementation of the opening policy. When $GC > 0$ the equity of urban residents to enjoy green space is reduced; when $GC < 0$ residents enjoy more equity of green space after opening the community; and when $GC = 0$ equity remains unchanged.

2. Simulating Equity-Oriented Urban Green Space Optimization Scenarios

Upon completion of the preliminary calculations, genetic algorithms can be leveraged to explore optimization solutions. As described in the opinions, opening the existing gated communities is a "step-by-step" process. There are different priorities for dismantling fences, which will be performed at different stages. Therefore, quantitative calculation is necessary to determine which gated communities should be preferentially opened at each stage (He et al. 2018). Based on the principle of space accessible distance, this paper selected 10 km as the distance that residents will travel to reach a green space, and divided it into the following: easy reach distance (≤1 km), difficult walking distance (2.5 km), transportation-accessible distance (5 km), and hard-to-reach distance (10 km). These four distance thresholds were regarded as different stages in the community opening process. A genetic algorithm simulates the biological mechanism of survival of the fittest, and optimizes and evaluates multiple candidates at the same time. The algorithm includes species, chromosomes, genes, and other elements, as well as crossover, mutation, selection, and other operations. A relationship between the genetic algorithm elements and space was established, and the chromosomes were mapped to the community opening scheme. The gene was characterized as each community, and there were two statuses, gated or opened, where gated was set to 0 and opened to 1. The genetic algorithm evaluated the optimization of each iteration using the fitness function after self-mutation, crossover, and selection operations (Fig. 3.12).

The rules for each operator and fitness function were as follows:

1. Mutation as the Primary Operation. The mutation rule altered the status of a community. A priority probability of opening for each community was set, which was determined according to the following equation:

$$P_{o\_i} = w_1 \times \mathrm{RoadD}_i + w_2 \times \mathrm{CGreenA}_i + w_3 \times \mathrm{CFA}_i$$

where P_{o_i} is the priority of the opening of the i-th community, and RoadD_i, $\mathrm{CGreenA}_i$, and CFA_i represent the standardized road kernel density, the green area in a community, and the footprint area occupied by the community, respectively. The latter three indicators represented the primary roles of residential units in urban road access, green space growth, and public space expansion, and were the main starting points for opening gated communities. w_1, w_2, and w_3 were set according to the advice of city planning department consultants.

The final mutation direction was determined by roulette. Assuming that the opening priority of a community was p, a random value r between 0 and 1 was generated, and the community would be opened if r was less than p, and it would remain closed when r was greater than or equal to p (Fig. 3.12a).

2. Crossover for Solution Recombination. The intelligent crossover operator randomly selected several pairs of chromosomes and two positions on each pair for exchange (Fig. 3.12b).
3. Selection-Driven Evolutionary Pressure. The main function of the selection rule was to screen dominant chromosomes, which was conducted by two levels of operators. The first-level selection operator selected the chromosome with higher fitness to serves as the next generation parent using the roulette rule after the mutation and crossover rules were complete. The second-level selection operator compared each chromosome before and after mutation or crossover, retained the better one, and eliminated the other to ensure the search efficiency of the algorithm (Fig. 3.12c).
4. Followed by Fitness Function Evaluation. The fitness function is a mathematical expression of the optimization objectives. In this study, the improvement of green accessibility (the average increase in MAF) and increased green equity (the decrease in GC value) were the main goals. Therefore, the fitness function was set as follows:

The optimality of a community opening scheme increases with the ratio of residents' average green accessibility to the Gini index of green accessibility in the region.

$$F = \frac{\mathrm{MAF}}{\mathrm{GC}}$$

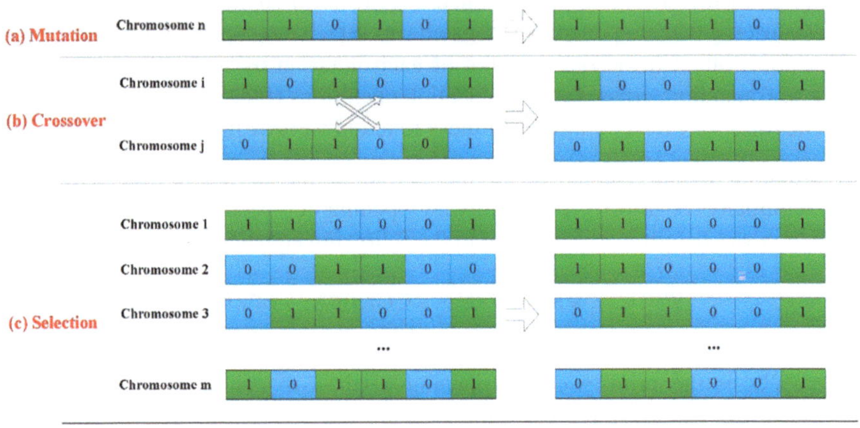

Fig. 3.12 Operational rules of genetic algorithm. (**a**) Mutation, (**b**) crossover, (**c**) selection

where MAF is the average green accessibility by the residents, and GC is the Gini index for green accessibility in the region. Evidently, our optimization goal was a high value of F.

3. Changes in Accessibility and Equity: Pre- and Post-Optimization

Following computational analysis, this study reveals the spatial clustering patterns of high green accessibility communities across Beijing's urban districts as distance thresholds increase. It further quantifies accessibility changes under gated versus open community scenarios, examines the correlation between green space allocation rationality and housing prices at varying distance thresholds, and identifies optimal community opening schemes.

In a prior study applying identical methodology in Wuhan, results revealed significant differences in green accessibility across distance thresholds before and after community opening (Fig. 3.13). Post-opening improvements were observed in overall green accessibility rates and residents' mean green accessibility, yet these averages masked substantial inter-community disparities. Crucially, green equity showed no improvement within 3.6 km thresholds due to community opening, but demonstrated optimization beyond 3.6 km. Given that most residents engage in daily leisure activities within shorter ranges, optimizing community opening sequences to enhance green equity at close-range thresholds (e.g., 1 and 2.5 km) is imperative. In the Wuhan study, the genetic algorithm optimization determined community opening schemes within the 1 and 2.5 km thresholds (Fig. 3.14).

The results revealed that 1423 and 1448 communities should be, respectively, opened within these two distance thresholds, accounting for approximately 66.3 and 67.5% of the total number of communities. Compared to the closed state, the optimized plan increased the average green accessibility by 27.65% and 31.22% for the two thresholds, respectively, and generally achieved the goal of not only increasing the opportunity for residents to enjoy green spaces, but also eliminating the original huge difference in this opportunity. Housing prices in the communities identified for opening within the 1 and 2.5 km threshold optimization scheme were basically the

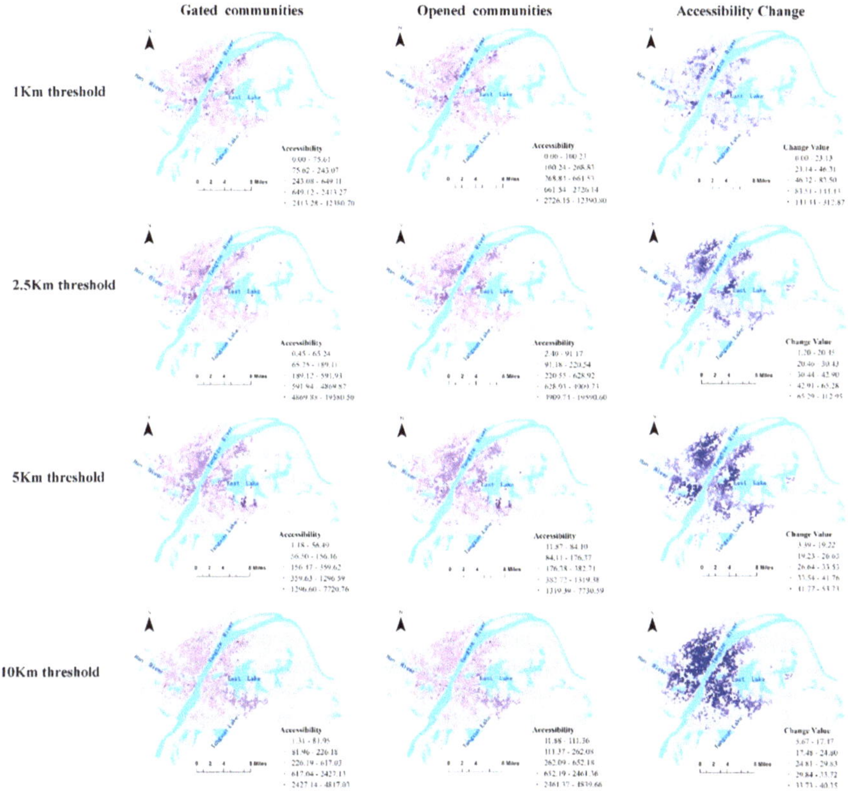

Fig. 3.13 Green space accessibility within four distance thresholds before and after community opening in Wuhan

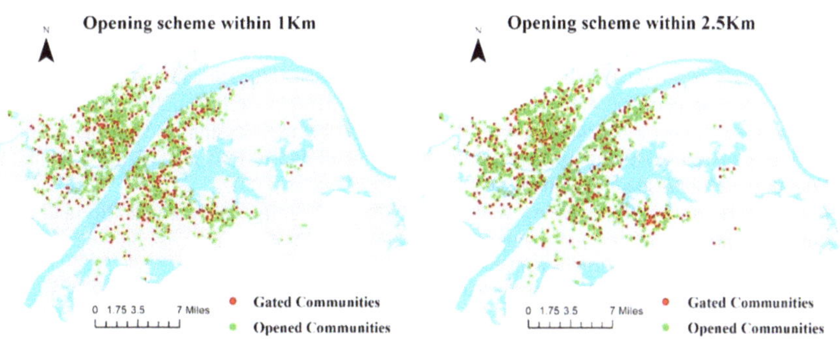

Fig. 3.14 Community opening schemes within 1 and 2.5 km distance thresholds in Wuhan

same as those for communities that continued to be closed. High- and low-price communities were balanced within the selection. In the future, for closed communities that need to be opened, the government should consider giving these communities additional economic compensation for the losses caused by opening the community fences, such as a loss of security and increased property maintenance costs.

The research framework and methodologies applied in Wuhan can be effectively extended to green space equity studies in Beijing. By employing the 2SFCA method and Gini index, it becomes feasible to quantify changes in green accessibility equity before and after community opening policies. Further integration of genetic algorithms enables optimized sequencing of community opening schemes. These innovations collectively advance the quantification of spatial relationships between residents and green resources, ultimately enhancing the equity of green space allocation.

References

Atkinson R, Blandy S (2005) Introduction: international perspectives on the new enclavism and the rise of gated communities. Hous Stud 20(2):177–186. https://doi.org/10.1080/0267303042000331718

Dai D (2010) Black residential segregation, disparities in spatial access to health care facilities, and late-stage breast cancer diagnosis in metropolitan Detroit. Health Place 16(5):1038–1052. https://doi.org/10.1016/j.healthplace.2010.06.012

Gieryn TF (2000) A space for place in sociology. Annu Rev Sociol 26(1):463–496. https://doi.org/10.1146/annurev.soc.26.1.463

He Q, Tan S, Yin C, Zhou M (2018) Collaborative optimization of rural residential land consolidation and urban construction land expansion: a case study of Huangpi in Wuhan, China. Comput Environ Urban Syst 74:218–228. https://doi.org/10.1016/j.compenvurbsys.2018.11.005

Lang RE, Danielsen KA (1997) Gated communities in America: walling out the world? Hous Policy Debate 8(4):867–899. https://doi.org/10.1080/10511482.1997.9521281

Lee G, Hong I (2013) Measuring spatial accessibility in the context of spatial disparity between demand and supply of urban park service. Landsc Urban Plan 119:85–90. https://doi.org/10.1016/j.landurbplan.2013.07.001

Li X, Zhang C, Li W, Ricard R, Meng Q, Zhang W (2015) Assessing street-level urban greenery using Google Street View and a modified green view index. Urban For Urban Green 14(3):675–685. https://doi.org/10.1016/j.ufug.2015.06.006

Luo W, Whippo T (2012) Variable catchment sizes for the two-step floating catchment area (2SFCA) method. Health Place 18(4):789–795. https://doi.org/10.1016/j.healthplace.2012.04.002

Piovani D, Arcaute E, Uchoa G, Wilson A, Batty M (2018) Measuring accessibility using gravity and radiation models. R Soc Open Sci 5(9):171668–171668. https://doi.org/10.1098/rsos.171668

Rigolon A (2016) A complex landscape of inequity in access to urban parks: a literature review. Landsc Urban Plan 153:160–169. https://doi.org/10.1016/j.landurbplan.2016.05.017

Su S, Li D, Yu X, Zhang Z, Zhang Q, Xiao R, Zhi J, Wu J (2011) Assessing land ecological security in shanghai (china) based on catastrophe theory. Stoch Env Res Risk A 25(6):737–746. https://doi.org/10.1007/s00477-011-0457-9

Su S, Zhou H, Xu M, Ru H, Wang W, Weng M (2018) Auditing street walkability and associated social inequalities for planning implications. J Transp Geogr 74:62–76. https://doi.org/10.1016/j.jtrangeo.2018.11.003

Webster C, Glasze G, Frantz K (2002) The global spread of gated communities. Environ Plann B Plann Des 29(3):315–320. https://doi.org/10.1068/b12926

Wu J, He Q, Chen Y, Lin J, Wang S (2018) Dismantling the fence for social justice? Evidence based on the inequity of urban green space accessibility in the central urban area of Beijing. Environ Plann B Urban Anal City Sci 47(4):626–644. https://doi.org/10.1177/2399808318793139

Wu J, Feng Z, Peng Y, Liu Q, He Q (2019) Neglected green street landscapes: a re-evaluation method of green justice. Urban For Urban Green 41:344–353. https://doi.org/10.1016/j.ufug.2019.05.004

Wu J, Peng Y, Liu P, Weng Y, Lin J (2022) Is the green inequality overestimated? Quality reevaluation of green space accessibility. Cities 130:103871. https://doi.org/10.1016/j.cities.2022.103871

Yu W (2017) Assessing the implications of the recent community opening policy on the street centrality in China: a GIS-based method and case study. Appl Geogr 89:61–76. https://doi.org/10.1016/j.apgeog.2017.10.008

Chapter 4
Research Perspective on Individual-Scale Urban Green Space and Equity

4.1 Urban Green Equity and Individual Mobility Patterns

As established in the preceding chapter, static exposure contexts predominantly centered on residential locations exist alongside dynamic exposure environments shaped by individual mobility. The divergence between static geographical contexts and dynamic exposure environments has been empirically validated across multiple disciplines (Burgoine and Monsivais 2013; Kwan 2022). This suggests that workplace and mobility environments may play a critical role in determining individuals' green exposure status (Park and Kwan 2016; Yu et al. 2018). Empirical studies reveal that travelers exhibit willingness to ride longer distances (e.g., commuting or school runs destination-oriented journeys typically prioritizing shortest-path routing) compared to recreational trips, to gain green surroundings on their route. This phenomenon may be related to the compensatory use of green spaces (Hall and Page 2014). The compensation hypothesis states that a group lacking UGSs around its place of residence may actively come in contact with UGSs in non-neighborhood environments to compensate for the shortcoming (Vedel et al. 2017). Therefore, in the present research, we believe that people may be affected by a static green environment (i.e., residence and workplace) and produce corresponding "compensatory" green exposure in the travel environment. The daily mobility will alleviate the green inequality caused by the static green space environment difference to some extent.

Under the background of green inequality and UGCoP, can daily mobility alleviate green inequality in living and working environment? We assume that people who lack green exposure will be exposed to more greenness in their daily mobility out of compensation psychology to narrow the greenness gap between their static geographical environment (i.e., workplace and residence).

This study investigates the living and working green environments of 554 respondents and their dynamic green exposure during active travel (AT). According to the

J. Wu, *Urban Green Space Equity in China*, Urban Sustainability,
https://doi.org/10.1007/978-981-95-5771-4_4

three dimensions of static green environmental indicators (i.e., availability, visibility, and accessibility), we divide the entire sample into sereval pairs in a high green index group and low green index group. By comparing whether significant differences in dynamic green exposure exist between the two groups, we test whether inequality in green exposure exists (Fig. 4.1).

4.1.1 Measurement Methods for Individual-Level Static and Dynamic Green Exposure

With GPS positioning, a travel log, and a questionnaire, a week-long survey of Beijing residents' daily travels was conducted in eight waves in 2012. With the assistance of the neighborhood committee of each community and human resource department of a company, 0.5–1% of the residents were randomly selected from 46 communities or companies on Shangdi-Qinghe street (Fig. 4.2). A total of 791 participants were recruited, and 709 respondents completed the survey, with a completion rate of 89.6%. The dataset documents basic attributes of each respondent, including gender, educational attainment, annual income, job-housing distance, and geocoordinates of residence/workplace. The respondents were required to carry a GPS tracking device for 1 week and record and upload their spatial and temporal coordinates every 30 seconds (Fig. 4.3). A travel log recorded details of each trip, such as vehicle type, duration, and origin and destination (home, workplace, or

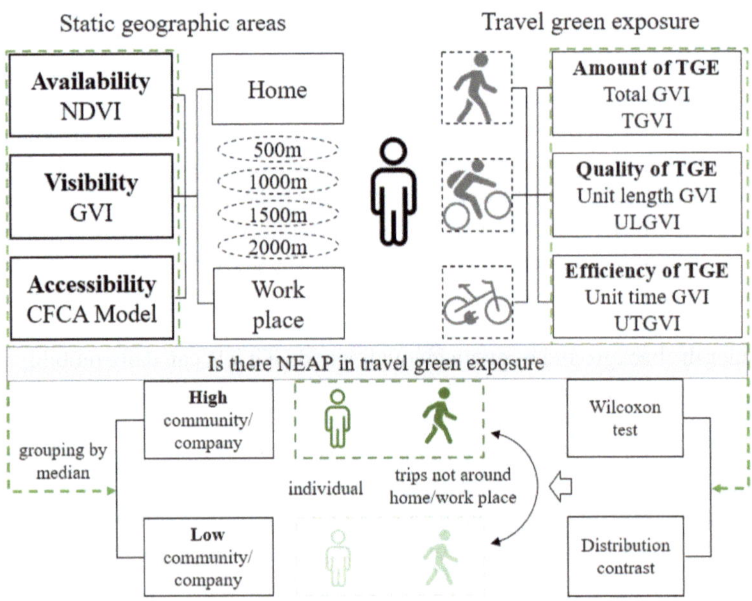

Fig. 4.1 Research design for static versus dynamic green exposure assessment

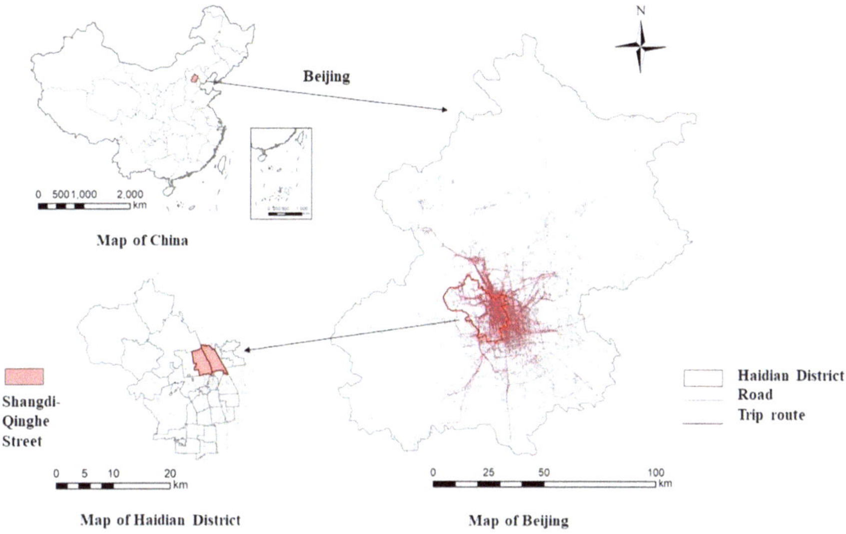

Fig. 4.2 Sample source location map

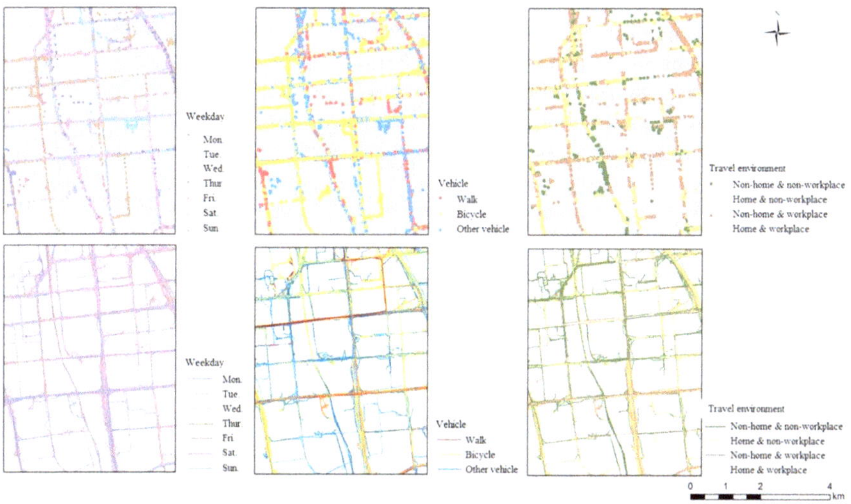

Fig. 4.3 GPS points and travel trajectory (GPS points are above, travel trajectories are below)

other). Excluding records with missing or invalid values, the valid sample consisted of 9802 trips, including 2156 ATs from 554 individuals.

1. Dynamic Green Exposure Measurement

Dynamic exposure predominantly occurs during active transport (AT) modes such as walking and cycling. As a transportation modality characterized by direct street greening contact and physical activity generation, AT constitutes a primary pathway for realizing green space health effects. Concurrently, green spaces serve as critical components of cycling/walking-friendly streets. Research on green space preferences aims to increase visitation frequency and/or physical activity levels, with core mobility patterns encompassing walking, jogging, and cycling. First, the street view big data are sampled, and machine learning is used to calculate the GVI of each road. Then, the respondents' perception of green spaces is described using the total GVI (TGVI), unit length GVI (ULGVI), and unit time GVI (UTGVI) to represent the total amount, quality, and efficiency of dynamic green exposure.

The street view sampling process is as follows. First, based on the road network data obtained from Amap (https://www.amap.com/), approximately 100,000 sampling points are generated at an interval of 100 m. Images are collected at 60-degree intervals in the horizontal direction to cover a 360-degree street view, that is, six images at each sampling point. After sampling, 500,000 images are obtained after the manual cleaning of the captured images under the condition of defoliation or autumn color change. Deep learning algorithms are used for the batch recognition of vegetation. A training set from Cityscapes dataset (https://www.cityscapes-dataset.com/) and an open-source urban streetscape image set with fine semantic annotations are provided, including 2975 images for training, 500 for verification, and 1525 for testing. TensorFlow and DeepLabv3 models are used in the training, and the parameters are set as follows: the batch size is 4, the initial learning rate is 0.001, the train crop size of the neural network is adjusted to the original size of the Cityscapes images, and the other parameters are trained according to the default parameters of DeepLabv3. These hyperparameters play a vital role in the convergence rate and accuracy of deep learning and are limited by the performance of computers (Smith 2018). The mIOU (an evaluation index of semantic segmentation accuracy) using 500 fine-annotation images from the validation set of Cityscapes is 75.79%. After the vegetation (including branches, leaves, and trunks) in each image is identified, the pixel proportion is calculated as the GVI value. The GVI value of each road is the average value of the GVI of all the images sampled on the road. The specific process of acquisition, machine learning, and calculation is shown in Fig. 4.4.

After the road network with the GVI is obtained, the GVI of each trip is calculated. The matching mode of the travel path and road is not through the buffer but through the following procedure. First, for the two adjacent GPS points P_{ijk} and $P_{ij(k+1)}$ in the i-th trip of the j-th individual, their GVI value is recorded as GP_{ijk} and $GP_{ij(k+1)}$, which derive from the GVI value of the nearest road. Moreover, Line_{ijk} is the segment between the two points, and its GVI value GL_{ijk} is:

$$GL_{ijk} = \left(GP_{ijk} + GP_{ij(k+1)}\right) / 2.$$

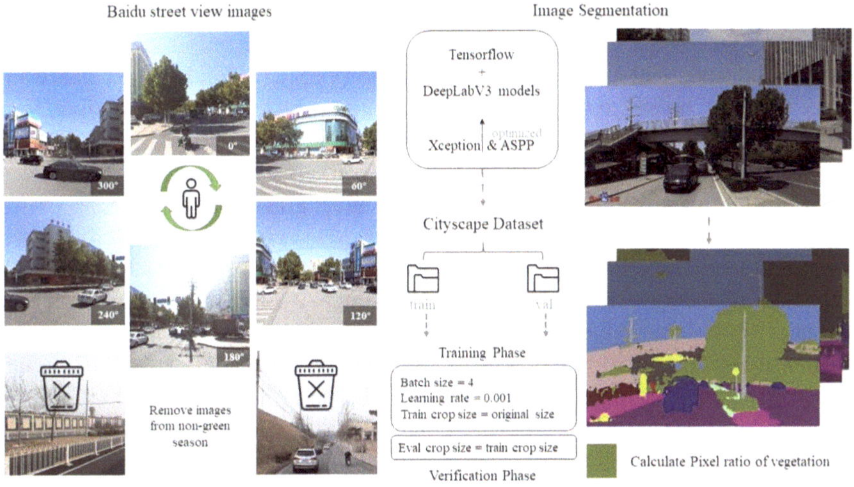

Fig. 4.4 Street view image acquisition, machine learning and GVI calculation process

The TGVI of Line_{ijk} is calculated as:

$$\text{TGL}_{ijk} = GL_{ijk} \cdot L_{ijk},$$

where L_{ijk} is the length (m) of Line_{ijk}. Thus, the TGVI for a trip is:

$$\text{TGVI}_{ij} = \sum_k \text{TGL}_{ijk}.$$

And the total GVI of all ATs for individual j over 1 week, denoted TGVI_j, is defined as:

$$\text{TGVI}_j = \sum_i \text{TGVI}_{ij}.$$

The unit-length GVI encountered by individual j, denoted ULGVI_j, is:

$$\text{ULGVI}_j = \text{TGVI}_j / \sum_i \sum_k L_{ijk}.$$

The unit-time GVI encountered by individual j, denoted UTGVI_j, is:

$$\text{UTGVI}_j = \text{TGVI}_j / \sum_i \sum_k T_{ijk},$$

where T_{ijk} is the time of Line_{ijk} went through, and $\sum_i \sum_k T_{ijk}$ is the total time (s) of the weekly AT of the j-th individual. The TGVI reflects the total amount of green landscape perceived by an individual in AT, and the difference in the TGVI may derive from the difference in the ULGVI or UTGVI.

2. Static Green Exposure Measurement

The static green exposure geographical areas are the four buffer zones (i.e., 500, 1000, 1500, and 2000 m) around the coordinates of each respondent's residence and workplace. According to the above content, static green exposure is described from three dimensions, namely, availability, visibility, and accessibility. The corresponding indices are the NDVI and average GVI of the roads within the buffer, and accessibility is based on the UGSs and surrounding street greenery.

This study utilized Landsat 8 imagery (May 2013 acquisition) with cloud cover <0.1. Following radiometric calibration and atmospheric correction, the NDVI was calculated as the ratio of the difference to the sum between near-infrared (NIR) and red band reflectance values. The GVI is the mean value of the GVI of all the roads within the buffer. In Sect. 3.1, how to calculate the GVI of all the roads through street views and machine learning is described. Therefore, the average GVI of the roads in the buffer zone is the average GVI weighted by the road length (Fig. 4.5).

The calculation of green space accessibility was based on adjustments to the CFCA model, with the specific procedure as follows:

First, based on the data of the second land survey in Beijing, 2042 UGSs labeled "parks and green spaces" and larger than 1 hm^2 are extracted. Based on the code for the design of public parks (GB 51192-2016) in China, the UGSs are divided into five categories according to area size. The catchment size is estimated with the area and travel mode. In this study, only walking and cycling are considered. The details are shown in Table 4.1.

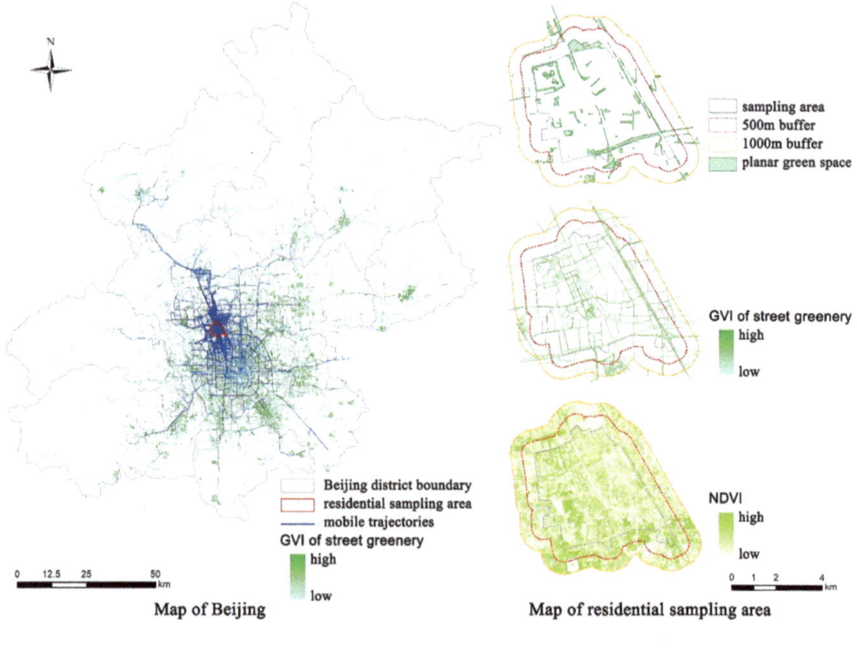

Fig. 4.5 Static green space distribution map

Table 4.1 Catchment area sizes corresponding to various scales and travel modes

Area (hm^2)	Number	%	Time (min)	(km)	(km)
≤2	1091	53.43	5	1.25	0.5
2–5	589	28.84	10	2.5	1
5–10	209	10.24	20	5	2
10–40	137	6.71	30	7.5	3
≥40	16	0.78	50	12.5	5

Second, UGS attraction is calculated. Each UGS has two attraction indicators, namely, area and the average GVI. The GVI attraction of walking and cycling should be calculated. The walking situation is taken as an example. First, the area of each UGS and average value of the surrounding GVI are calculated, and the specific calculation range is shown in Table 4.1. Through the use of min–max normalization to achieve nondimensionalization, the size and GVI of the UGSs are normalized to Area_s and GVI_Walking_s. The attraction of the i-th UGS in the walking mode is:

$$\text{G\_Walking}_i = \beta_1 \text{Area}_{si} + \beta_2 \text{GVI\_Walking}_{si},$$

where β_1 and β_2 represent the weights, and the total of β_1 and β_2 equals 1. The weight coefficient is calculated with the entropy weight method. For Area_s, the information entropy is:

$$E_{\text{area}} = \begin{cases} -\ln(n)^{-1} \sum_{i=1}^{n} p_i \ln(p_i), & p_i \geq 0 \\ 0, & p_i = 0 \end{cases},$$

where p_i is the proportion of the i-th UGS area in the total UGS area, as follows:

$$p_i = \text{Area}_{si} / \sum_{i=1} \text{Area}_{si}.$$

The information entropy ($E_{\text{GVI\_Walking}}$) of GVI_Walking_s can be obtained in the same way:

$$\beta_1 = \frac{1 - E_{\text{area}}}{2 - E_{\text{area}} - E_{\text{GVI\_Walking}}},$$

$$\beta_2 = 1 - \beta_1.$$

Similarly, UGS attraction in the cycling mode (G_Biking_i) is calculated. Based on the principle that attraction decreases with distance, the accessibility of each residential or working point is calculated. The accessibility of the j-th individual's residence is taken as an example, as follows:

$$\text{Access\_home}_j = \sum_{i \in \{d_{ij} \le d_0\}} \text{G\_Walking}_i D(d_{ij}, d_0) + \sum_{i \in \{d_{ij} \le d_0\}} \text{G\_Biking}_i D(d_{ij}, d_0),$$

where

$$D(d_{ij}, d_0) = \begin{cases} \dfrac{e^{-\frac{1}{2} \times \frac{d_{ij}}{d_0}} - e^{-\frac{1}{2}}}{1 - e^{-1/2}}, & d_{ij} < d_0, \\ 0, & d_{ij} \ge d_0 \end{cases}$$

where d_{ij} is the distance between the i-th UGS and j-th individual's residence, and d_0 refers to Table 4.1. The calculation in this step considers the walking and cycling scenarios. In the same way, the green accessibility of the workplace is calculated.

To test whether individuals generate compensatory green exposure in mobility to alleviate the greenness gap between their static environments (i.e., residence and workplace). On the basis of the above measurements of static and dynamic green exposure, we took the median of NDVI, GVI, and accessibility at each respondent's residence and workplace in four buffer zones as the standard line and divided the full sample into 24 pairs of high and low static green exposure (277 people in each group). After comparing the mean and difference of dynamic green exposure of each group pair, we analyzed the above problems using the individual and behavior basic analysis units.

4.1.2 Individual-Level Differences in Green Exposure

1. Differences Between Static and Dynamic Green Exposure

We show the difference in the frequency distribution of the group pairs according to accessibility and the GVI in the 500-m buffer (static green exposure indicators are standardized) in Fig. 4.6. We assumed that people who lack green exposure in their living or working environment would actively seek more greenness when traveling. Yet, our results showed that at the individual level, people living in a highly green environment also maintain a higher dynamic green exposure. Figure 4.6 demonstrate two main aspects, that is, increased zero values in the dynamic green exposure of the residents or employees in low-accessibility communities or companies and the high frequency of the high dynamic green exposure of the residents or employees in high-accessibility communities or companies.

Therefore, the weekly dynamic green exposure of the individuals may be polarized given the green space environments of their residences or workplace, and daily mobility does not alleviate green inequality. When accessibility and number of green spaces are high, individuals will maintain high dynamic green exposure in their non-neighborhood trips. First, a satisfactory neighborhood green space

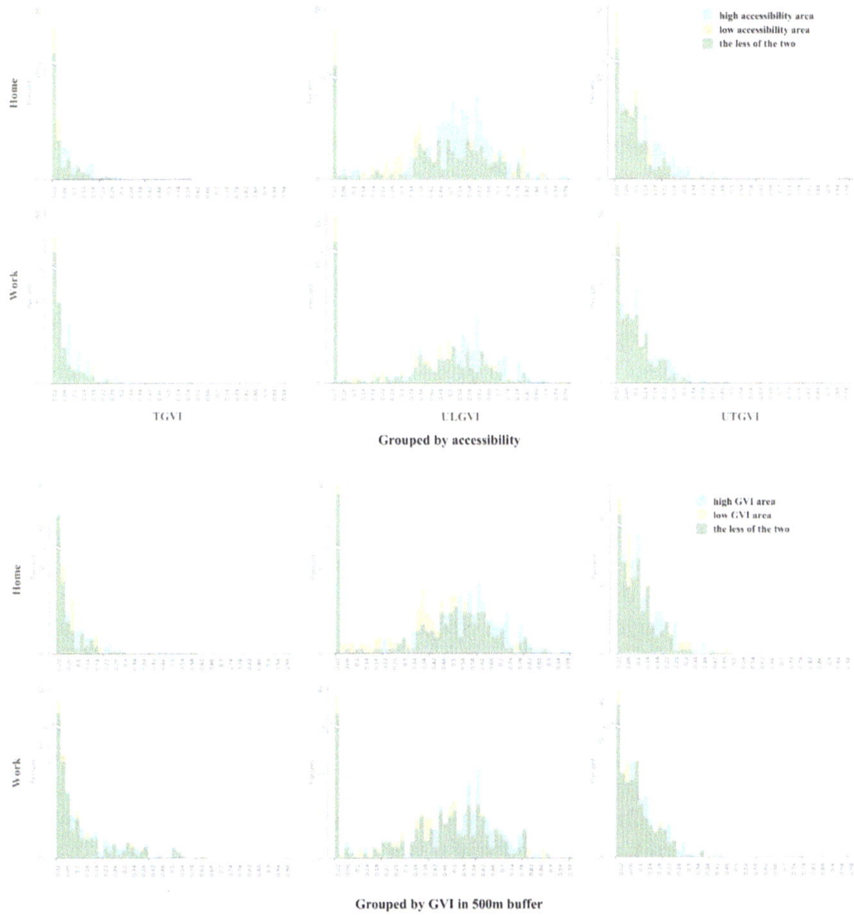

Fig. 4.6 Comparison of frequency distribution of individual travel green exposure in 1 week

environment may encourage residents to actively seek greenness in their travels; Second, the choice of living and travel environments involves self-selection, which is affected by personal preferences. Residents who prefer AT tend to use considerable green spaces, thereby widening the gap between AT and green exposure. However, the results at the travel level indicated that employees who are at a disadvantage in green would be exposed to higher-quality green spaces in non-workplace neighborhood travel. In the workplace context, the compensation behavior of green exposure is more likely to occur, although such compensation behavior may not change the current situation of their vulnerable green status. The contradiction between the individual and travel levels means that employees who are at a green disadvantage may be exposed to high-quality green spaces during their non-workplace neighborhood travels. Groups that experience considerable pressure tend to choose UGSs with natural elements. Therefore, people's compensation behavior for green exposure in work-related travel may be related to the restorative effect of plant landscapes on work stress.

2. Disparities in Green Exposure Between Residential and Workplace Environments

Results of mean comparisons and intergroup difference tests revealed that: (1) Significant differences in dynamic green exposure during non-neighborhood trips exist between residents of high-accessibility and low-accessibility communities. Specifically, residents in high-accessibility communities exhibited higher travel green exposure levels. (2) Higher community NDVI is moderately associated with enhanced non-neighborhood dynamic green exposure quality. Non-workplace trips were influenced by small-scale NDVI, while low-NDVI workplace environments motivated respondents to engage in trips with higher dynamic green exposure. (3) Higher GVI within 500 m of communities correlated with increased TGVI during non-neighborhood trips, achieved by opting for higher UTGVI paths. For workplace GVI, although high-GVI workers showed lower UTGVI and ULGVI, they maintained relatively higher TGVI.

In the results of travel level, green travel exposure in residential context is polarization, while that in work context is compensation. There may be two reasons for this difference. First, use of green space initiatives differs between the living environment and working environment. Neighborhood green spaces are important for residents for engaging in leisure and social activities. However, those who work often "forget" or "have no time" to use green spaces owing to their busy schedules. In this study, the number of home-based leisure and entertainment AT is approximately 2.6 times that of work-based AT, which further shows that people's motivation to seek green spaces in their working environment is insufficient. Second, greenness may have different impact mechanisms on individuals' travel behavior in their residences, workplace, and actual travel space, resulting in different dynamic green exposures. Prior studies found that the attractiveness of workplace greenness to individuals' physical activity is not as good as that of their residences and determined a negative correlation between greenness and PA frequency around the workplace. Therefore, in the workplace context, the influence of urbanization density, traffic convenience, and land-use mix on individuals' behavior is greater than that of greenness. The above studies also emphasized the links between living, working, and other environments and the possibility of compensatory behavior. A busy and high proportion of utilitarian behaviors lead to inefficient use of green space around the workplace. Consequently, people may tend to use green space of residence and travel space to compensate for the gap brought by working green space environment.

Synthesizing the above findings: First, at the individual level, disparities in greenspace accessibility and street greening quality at residential or workplace locations contribute to green injustice in dynamic green exposure. Second, a contrary pattern emerges for trips beyond residential and workplace peripheries. Specifically, residents from communities with higher greenspace quantities experience higher-quality street greening during mobility, whereas employees in workplaces with poor greenspace quantity and street greening quality exhibit more efficient compensatory exposure to superior urban greenspace. Paradoxically, workplace green environments inversely influence green exposure during non-workplace trips, indicating that overlooking workplace contexts and their impact on dynamic exposure may lead to assessment bias in green injustice evaluation.

4.2 Urban Green Equity and Neighborhood Effect Polarization

With the in-depth study of neighborhood effects, researchers gradually realized that people's activities in the residential neighborhood environment are very limited. The uncertain geographic context problem (Kwan 2012), the selective daily mobility bias (Chaix et al. 2013; Perchoux et al. 2016), and other issues point out that when studying the relationship between environment and residents' attitudes, behaviors, and health, ignoring mobility and nonresidential neighborhood environment will lead to the inaccurate measurement of the actual environment exposure of individuals, and the neighborhood effect of environmental factors will be underestimated or overestimated. These errors have been confirmed in a variety of mobility-dependent environmental exposures, such as food environment, air pollution, and green space. Kwan further pointed out that environmental exposure might converge to average level due to individual mobility, and generalized this phenomenon as the neighborhood effect averaging problem (NEAP) based on the results of several empirical studies (Park and Kwan 2016; Shafran-Nathan et al. 2016; Yu et al. 2018).

NEAP, however, might be limited by environmental types, spatial scale and place, and special groups. In terms of environmental types, NEAP should mainly exist in environmental factors that are highly dependent on mobility rather than more related to residential areas (e.g., collective efficiency). Yet, the current NEAP is only found in exposure to air pollution and traffic congestion (Kim and Kwan 2018, 2021a, b). No evidence shows that NEAP is equally applicable to other mobility-dependent environments (e.g., food environment, green space environment, medical environment). With regard to the spatial scale and place, evidence was found mostly in Western cities. The habits and behavioral patterns of different urban residents might lead to a change in the NEAP pattern and affect the effectiveness of the NEAP owing to the change in the activity places (Tan et al. 2020). Moreover, when the pollution concentration does not change much in space, NEAP will be weak or nonexistent (Kim and Kwan 2021a). Regarding groups, compared with high-income, employed, young, and male groups with strong mobility, low-income groups, elderly people, and those restricted by housework will find participating in NEAP difficult because of limited daily activities. In addition, research showed that the characteristics of mixing and segregation of racial exposure vary from race to race. This series of evidence implies that mobility-based environmental exposure does not always tend to be average (or converged).

4.2.1 *The Neighborhood Effect Polarization Problem in Green Exposure*

The neighborhood effect polarization problem (NEPP) is another possible situation for the neighborhood effect bias of mobility-dependent environment: The overall trend of mobile exposure is more polarized than that of residential exposure. Specifically, people with high residential exposure will have higher mobile exposure, thereby making the neighborhood effect of the residential environment underestimated. First, different environmental factors have completely different mechanisms of avoiding harm and approaching benefit (Bresnahan et al. 1997; Ulrich 1986). Second, environmental demands might exist in the self-selection process of neighborhood and travel space (Kim and Kwan 2021a). Finally, NEAP is limited by mobility restriction and social isolation. As a typical representative of positive environmental factors, we infer that green exposure is very likely to have NEPP because of the worldwide issue of green justice and its important position in the self-selection problem (Nawrath et al. 2019; Sarkar et al. 2015; Zhang et al. 2012).

1. Environment Driven Behavior: Approaching or Avoiding
At present, the evidence for NEAP mainly comes from pollution and traffic jam exposure, which are the factors that drive travel behavior by avoiding harm (Bresnahan et al. 1997). By contrast, people are driven by the benefit to approach other environmental factors (e.g., green space). Psychological research found that people tend to approach positive stimulus and avoid negative stimulus faster than they approach negative stimulus and avoid positive stimulus (Chen and Bargh 1999; Solarz 1960). Therefore, taking the two typical representative types of green space and pollution as examples, the environmental characteristics that drive residents to approach or avoid actively are attractive green space and highly polluted air, respectively.

Considering that city planners are committed to creating a positive environment and eliminating negative ones, however, residents are more likely to take the initiative to approach positive environmental characteristics. Particularly when the environmental characteristics are difficult to identify, or the spatial accuracy of the information is not enough to affect the route choice, residents might ignore the screening during their daily mobility. For example, residents will only have active avoidance behaviors when air quality alarms and health symptoms occur (Cutter and Neidell 2009; Liu et al. 2017; Zivin and Neidell 2009). Diversified food environments and healthy green spaces can easily attract travelers (Chen and Yang 2014; Krenn et al. 2014). Therefore, although a variety of environmental factors are considered in the selection process of neighborhood environment, the difference between the availability of environmental information and the behavior mechanism of approaching the positive and avoiding the negative allows the residents to more often approach the positive environment in the process of mobility.

2. Urban Green Space Demand: Compensation or Convergence
The characteristics of environmental demands are likewise pivotal in determining neighborhood effect trends of environmental exposure. The selection between

neighborhood and non-neighborhood environments may exhibit either complementary or convergent relationships. The aforementioned "compensatory behaviors" may thereby drive green exposure toward NEAP.

However, convergence demands for residential and nonresidential environments also exist, manifested in people exposed to similar environments under multiple geographic conditions. Regardless of their residential environment, people in non-residential environments are likely to visit places similar to their residential environments (Hall and Page 2014), thereby causing the NEPP. Although the compensation hypothesis has been around for a long time, the study of rejecting hypotheses has relatively more evidence. In addition, empirical evidence showed that residents have similarities in environmental exposure or demand in different environments. For example, the frequency of green space use is similar in three contexts, namely, residential environment, workplace, and other outdoor leisure places. Negative environmental factors, such as noise and pollution, will seriously affect people's health, travel satisfaction, and residential happiness (Kestens et al. 2012). The planning and design of roads and the reduction of sensitivity might be insufficient to affect the process of travel and residential self-selection, however. This case limits the possibility of NEPP in some environmental exposures (Zenk et al. 2011).

3. Access Restriction: Resource Inequality and Social Differentiation

The NEAP is not consistently effective. The uneven distribution of environmental resources and social isolation caused by social differentiation will considerably limit individuals' mobility and could lead to changes in the type of neighborhood bias in some groups (Kim and Kwan 2021a, b; Ma et al. 2020). In NEAP-related research, high-income, employed, young, and male participants were more likely to have NEAP because of their high daily mobility. Coincidentally, studies in Beijing also found that low-income people with low daily mobility and limited travel outside their homes or outside the workplace and elderly people who are restricted by housework are likely to be the exception of NEAP. Thus, when the proportion of restricted mobility groups increases because of social differentiation, the neighborhood deviation is likely to change from NEAP to NEPP or will show an unbiased intermediate state. In addition to restricting mobility and leading to social isolation, the potential problems of social differentiation might be more serious than we thought. Environmental justice issues are found in various environmental exposure studies including differences in green space accessibility, food deserts, or differences in residential pollution exposure directly caused by differences in social and economic status, such as race, income, and age. Moreover, the uneven distribution of green space, food resources, policy interventions, and social isolation will further lead to the ongoing marginalization and even displacement of disadvantaged groups, resulting in green gentrification and food gentrification

Overall, researchers recognized that the NEAP could occur when considering daily mobility. Owing to the limitations of NEAP research, however, this article argues that this phenomenon might not be applicable to other environmental exposure types and cities. By contrast, owing to people's varying response behaviors to different environments, such as benefit-seeking and avoidance, compensation and convergence, and strong and weak mobility, we have reason to believe that NEPP will emerge as a form of neighborhood effect bias. Therefore, this study takes green exposure as a typical case and answers the following questions: (1) Does NEPP exist? (2) What neighborhood green space characteristics will lead to NEPP? (3) Is the convergence of people's environmental demands one of the causes of NEPP? This section will continue to utilize the Beijing sample established above for case analysis. By measuring their neighborhood green environments and travel-based green exposure, it demonstrates the existence of NEPP and discusses the necessary conditions for its emergence.

4.2.2 Residential and Mobility-Based Green Exposure Assessment Approaches

We calculated the residence-based and mobility-based green exposure separately. Moreover, the samples will be divided into green-advantaged groups and green-disadvantaged groups according to the residence-based green exposure indexes. To describe the total amount, quality, and efficiency of mobility-based green exposure (MGE), we measured the total GVI (TGVI), the unit length GVI (ULGVI), and the unit time GVI (UTGVI). The indicators of residence-based green exposure (RGE) include NDVI, GVI, and CFCA to describe the green space environment from three aspects: availability, visibility, and accessibility (Wu et al. 2020). The computational procedure aligns with the methodology detailed in the preceding section. We use the criteria importance through intercriteria correlation (CRITIC) method, which is an objective weighting method that considers contrast intensity and the conflicting character, to calculate and aggregate the weights into one comprehensive indicator according to the variability of five indicators and the correlation between the indicators. The subsample of the top 50 percent of the composite index is defined as the advantaged group, and the rest is the disadvantaged one. Figure 4.7 (left) shows the weight of each index. A significant correlation exists between the comprehensive index after weighted average and each index (Fig. 4.7, left). The

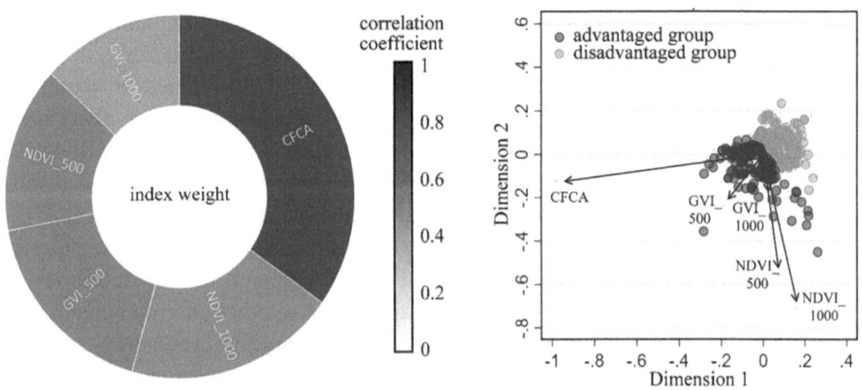

Fig. 4.7 Indicator weight pie chart (left) and principal component analysis biplot (right)

biplot shows that each subindex of the advantaged group is significantly higher than that of the disadvantaged group (Fig. 4.7, right). This comprehensive index can effectively distinguish advantaged and disadvantaged groups. In addition to green space factors, according to the 5Ds framework of Ewing and Cervero (2010), other built environment (BE) indicators that have a potential impact on green exposure will be regarded as control variables. The number of intersections, stations, and points of interest (POIs) is obtained from the open interface of Baidu map. The proportion of residential land (residence) and land-use mixing degree (LUM) are calculated from the data of the second national land survey at the end of 2010. The LUM is based on five types of land: residential, commercial services, public management and public services, transportation, and non-construction according to Shannon entropy. The average height of buildings is calculated from the average building height in the buffer zone.

4.2.3 Intrinsic Mechanisms of Green Exposure Neighborhood Effect Polarization

Following Wilcoxon tests, dispersion comparisons, and logit regression analysis, we investigated the intrinsic mechanisms driving neighborhood effect bias from both green space and individual perspectives. Figure 4.8a displays the forest plot of correlation coefficients (95% confidence interval [CI]) between three green environment indicators and mobility-based green exposure (MBE) in the full sample. The results demonstrate a convergence between residential and mobility-based green exposures: individuals residing in neighborhoods with low green accessibility are less likely to achieve compensatory MBE, while those in green-privileged communities not only consolidate their advantages through travel but also exhibit MBE levels that deviate further from the population mean. Green-advantaged groups show significantly higher MBE ($p < 0.05$), indicating substantial intergroup heterogeneity in nonresidential environmental exposure.

The frequency histogram in Fig. 4.8b compares the dispersion of green exposure indexes after mean normalization in nonzero samples, more intuitively proving the existence of NEPP, which is reflected in interval range and individual distribution. Among them, the order of standard deviation of each index from large to small is UTGVI (0.794) > TGVI (0.333) > ULGVI (0.236) > GVI_500 m (0.233) > GVI_1000 m (0.104). Compared with RGE, MGE has a larger discrete range and more individuals scattered in an interval far from the average.

Thus, there is NEPP in green exposure (measured by accessibility and availability); that is, considering daily mobility, people with high and low residential exposure will have higher and lower mobile exposure. There are two possible reasons for this polarization. First, although residents' preference for green space is common in travel, there could still be demand differences among different groups, which are discussed further in what follows. Second, there are group differences in the benefits of mobility. Mobility limits whether individuals choose to use green space, or

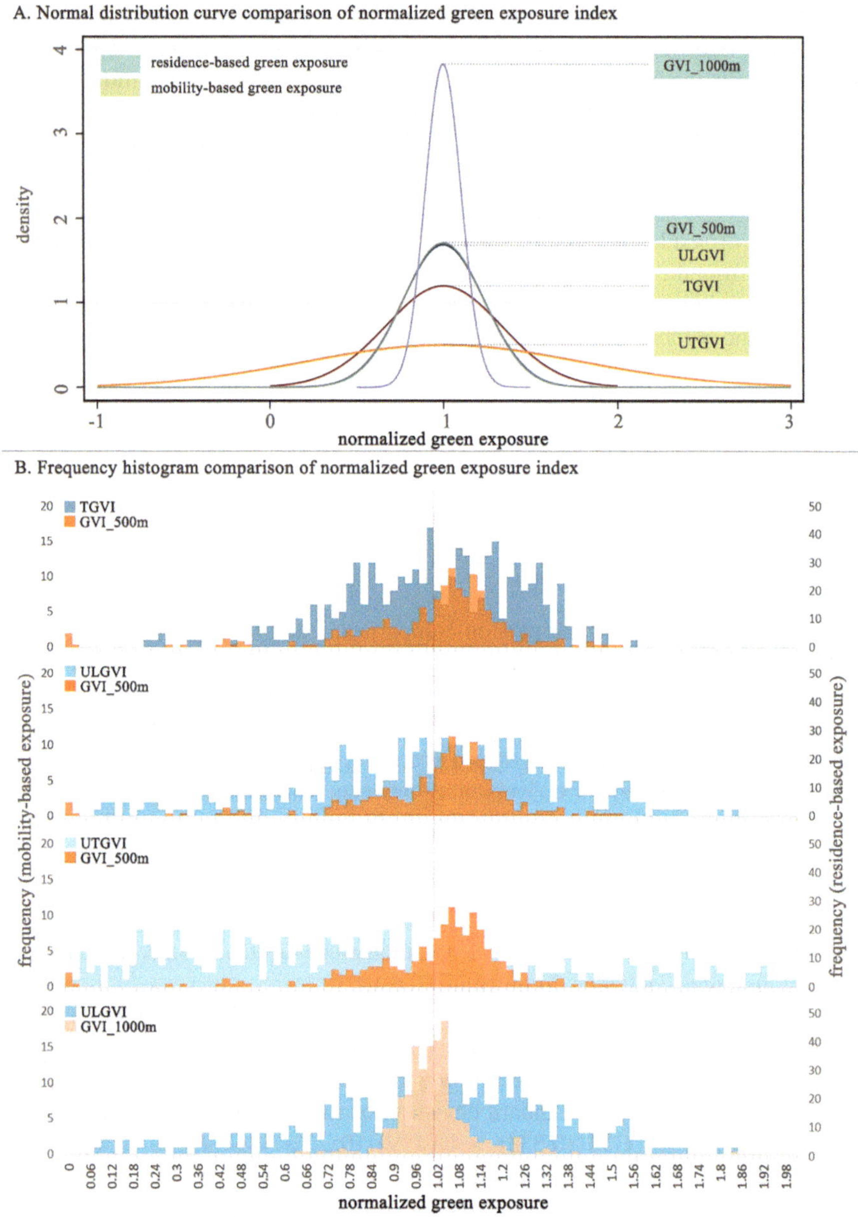

Fig. 4.8 Comparison of normal distribution curve and frequency histogram for standardized green exposure index. (**a**) Normal distribution curve comparison of normalized green exposure index, (**b**) frequency histogram comparison of normalized green exposure index

whether they choose to travel further on greener, better routes. As explained in the analysis framework, demand preference and mobility together lead to the emergence of polarization problems. Results reveal that residents easily choose environments with similar residential green space visibility during their travels, thereby resulting in NEPP. Therefore, the emergence of NEPP is related to differences in specific green space characteristics, and the neighborhood effects of green accessibility and visibility are likely to be underestimated when neglecting mobility. Previous studies on residents' green space landscape preference support this result. Accessibility is the main factor driving residents to contact urban green space, followed by vegetation coverage. In the case of better street greening, residents will appropriately extend the AT distance of social activities, so they also pay more attention to the green space environment around the residence. Therefore, increasing only the amount of urban green space has little effect on increasing green exposure, but increasing accessibility and street greening demands more attention from urban planners.

The convergence of environmental demands under different geographical backgrounds induces residents to visit similar environments, and the high demand for green space visibility of the advantaged group might become the leading cause of polarization. In fact, the study found that differences in green demands also extend to the workplace, and there is a strong positive correlation between the frequency of residents' participation in outdoor leisure activities and their participation in green space in the workplace. This result means that the green exposure of the neighborhood environment, workplace, and travel space has convergence, which further indicates that residence-based measures will underestimate the green space utility of actual contact. Previous research results support this argument: Although people tend to use more residential green space, RGE disadvantaged groups will not use more out of compensation psychology. Although the preceding results confirm that green spaces in a large neighborhood can encourage the MGE of disadvantaged groups, they cannot change the polarization trend in such groups. This outcome might be due to the mobile restrictions of disadvantaged groups, who experience difficulty narrowing the green exposure gap between themselves and advantaged groups through long-distance travel.

The existence of NEPP enables a more precise understanding of the environmental inequalities that individuals actually experience. When people's mobility motivations—such as the desire to seek more favorable environments—interact with constraints on their ability to move, the NEAP can break down and even evolve into NEPP. As a result, the dual disadvantages of environment and mobility created by social stratification further intensify environmental inequality. We must comprehensively consider how socioeconomic status affects individuals' environmental exposure and mobility under different temporal and spatial backgrounds from the perspective of distributional, procedural, and interactional justice. In addition, we must try to alleviate environmental inequality by restoring the effectiveness of the NEAP. On the one hand, although we obtain the polarization trend of green exposure in the overall mobile process, the encouraging effect of high-quality street greening on disadvantaged groups reveals the potential to solve environmental injustice by improving accessibility to residential green space and implementing widespread street greening (e.g., improving the urban greenway network with residential areas as nodes). On the other hand, the mobility restriction and social isolation of vulnerable groups, as reasons that could aggravate this problem of justice, are also aspects that need to be addressed by policy. Gradually opening the closed community and paying attention to the quality optimization of linear green space is one of the methods that can be explored. Therefore, future research must be expanded to include a wide range of environmental types, explore the internal mechanism of the formation or transformation of the NEAP and NEPP from the individual level, and seek strategies to avoid serious environmental justice issues.

4.3 Urban Green Equity and Gender Differences

In the study of urban green space equity, gender constitutes a pervasive yet often overlooked social determinant. Research indicates that women use urban green spaces at disproportionately lower rates relative to their demographic representation (Lu et al. 2024; Xu et al. 2023). However, women often place higher value than men on the importance of outdoor recreational activities. Additional studies reveal that homemakers, in particular, derive significant benefits from residential green spaces. When considering mobility issues, gender-based differences in spatiotemporal constraints, preferences, and household roles may lead to variations in travel schedules, modes of transportation, and destination choices (Kwan 2009, 2013; Matthews 2008). These differences further result in disparities in the quantity, quality, and efficiency of green space exposure during travel. Thus, distinct gender groups perceive and interact with urban green spaces in markedly different ways. Detailed analysis is essential to identify and address gender-related inequities and formulate appropriate planning policies.

4.3.1 Potential Determinants of Gender-Based Disparities in Green Exposure

In terms of spatiotemporal constraints, although the traditional idea that women are expected to undertake more housework has faded among younger urban middle and upper classes (Fan et al. 2011), while market goods and services have made housework more efficient and reduced the time necessary for household chores, women are still engaged in more housework and less entertainment, and often carry more competing responsibilities regarding family, social relations, and work (Konrad 2015; Silm et al. 2013). Accordingly, most in-home activities are typically fixed in space and time and act as "pegs" around which other activities are scheduled (Chidambaram and Scheiner 2021). As a result, time poverty disproportionately affects women. Moreover, it is likely more difficult for women than for men to visit more attractive urban green spaces frequently according to their own wishes or to enjoy the various benefits conveyed by green spaces. However, things may not be straightforward. Juggling employment and childcare may result in women's fewer flexible schedules (Kwan 2000; Ta et al. 2016), but a part of women's parenting responsibilities usually involves chauffeuring trips or taking children to parks and playgrounds (Scheiner 2016), which may, to some extent, advantage women over men. Particularly in the aspect of travel duration, although women may still have lower participation in long-distance trips and driving, they might have more complex trip patterns and longer trips near the residential area.

For choice preferences, many studies have reported differences in both physiological and psychological responses to greenness as well as selection and use of green space between social groups of different genders (Bolte et al. 2019; Kaczynski et al. 2009; Markevych et al. 2017; Schipperijn et al. 2010). Studies indicate women pay more attention to the natural environment and express more aesthetic pleasure in urban green space than men do, hoping to have safe and pleasant experiences in outdoor environments and to find social and recreational resources in them (Morris et al. 2019). On the other hand, women are more inclined to choose public spaces with sufficient light and better maintenance and management for activities (Crawford et al. 2007; Jones et al. 2009). Compared to men, women's sense of security is more strongly linked to the characteristics of the physical environment. Without effective management, an area of green space can become a disamenity because of a "fear of violence or crime" (Kimpton et al. 2016; Sreetheran and van den Bosch 2014). Further research is needed to verify whether this difference in choice preference leads women to shift travel patterns based on perceptions of green space and safety.

Finally, gender roles are related to roles in the family and society, which are reflected in the different ways in which men and women obtain rights and resources in family and social behaviors worldwide. Among them, in terms of family roles, men are more visible in the allocation of family resources and the distribution of rights to use family cars and other valued resources, and it is easier for men to travel long distances with the support of a car (Fagnani 1987; Hanson and Hanson 1980; Scheiner and Holz-Rau 2012). In terms of social roles, antiquated ideas left over from patriarchal society mean that occupations influencing the use of natural spaces, such as architecture and urban planning, have been and continue to be dominated by men (Schwanen et al. 2008). Even if the position and status of women have undergone considerable progress in recent decades, the gender gap in political engagement and decision-making still persists (Bell 2016). Therefore, urban development has long been influenced by principles and standards largely determined by men, and such urban planning decisions that ignore or are unfriendly to women may virtually compress women's travel distance and travel duration, affecting their exposure to green space.

4.3.2 *Examining Variations in Green Exposure Across Genders*

In summary, social groups based on gender often display distinctive patterns of daily activities and behavior patterns due to the differences in their spatiotemporal constraints, choice preferences, and gender roles, which in turn affect the quantity, quality, and efficiency of their exposure to green space (Fig. 4.9). Hence, we speculated that women would be more restricted than men in their daily life and travel so their range of activities in green space and the green exposure per unit time would be lower than men's. However, since women are likely more concerned with the aesthetic value and safety of the landscape, the quality of green exposure may bring about different results.

To mitigate biases stemming from gender and mobility dynamics, this study employs a gender-based approach to examine how variations in transportation modes, route choices, time constraints, and destination types during travel generate disparities in green exposure across gender groups, while investigating the mechanisms behind gender differentiation and its societal implications. We subsequently conduct an empirical analysis using Beijing as a case study, applying the urban green exposure metrics established earlier. The research first performs spatial autocorrelation tests to control for spatial dependence and multicollinearity. Independent

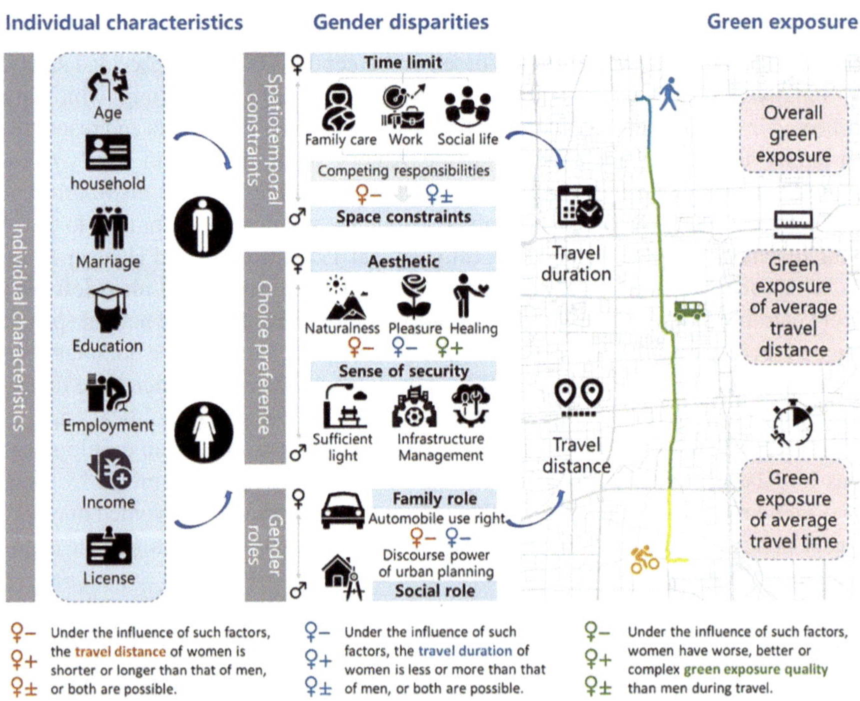

Fig. 4.9 Research framework for gender-based green exposure disparities

samples t-tests are then used to analyze gender differences in total travel duration, distance traveled, daily trip frequency, cumulative green exposure, green exposure per unit distance, and spatial quality efficiency. With cumulative green exposure, green exposure per unit distance, and green exposure per unit time as dependent variables, Least Significant Difference (LSD) post hoc testing further examines whether significant disparities exist in urban green exposure patterns across the full sample, male subsample, and female subsample.

4.3.3 Gender Differences in Green Exposure and Their Underlying Mechanisms

The multiple comparison results indicate that females exhibit significantly shorter average travel durations and distances than males, reflecting a tendency toward shorter-distance trips (Fig. 4.10). Significant gender disparities also exist in cumulative green exposure, with females demonstrating lower exposure levels. Per-trip

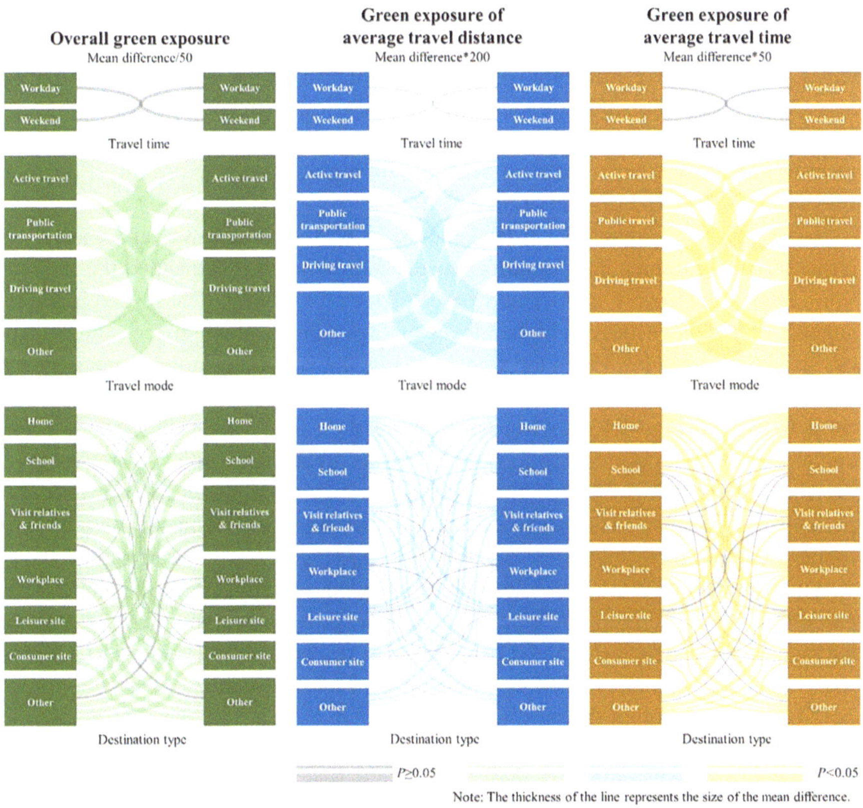

Fig. 4.10 Multiple testing of travel behavior characteristics in green exposure

visual green exposure for females is substantially reduced compared to males, and males gain 11% higher urban green space visibility per second than females. The results of multiple linear regression demonstrated differences in the overall green exposure and the green exposure of average travel distance. These differences were manifested in men enjoying higher levels of overall green exposure than women during each trip, while men's exposure to the green space per unit distance (m^{-1}) was lower. As far as individual characteristics were concerned, age, driver's license, marital status, employment, and income level were clearly some of the underlying mechanisms of gender differences in green exposure. With increasing age, individuals experienced lower overall green exposure in the sample of women and higher green exposure of average travel distance in the sample of men, though the trend was minimal. The overall green exposure, green exposure of average travel distance, and green exposure of average travel time obtained by women residents with a driver's license were likewise lower than those without a license, whereas among the samples of men, those with a driver's license had higher levels of green exposure for average travel distance and average travel time than men who had no license. Similarly, there is a complex connection between marital status and green exposure, as married men had lower levels of green exposure for average travel distance than those who were unmarried; married women, however, had higher levels of green exposure for average travel distance than those who were unmarried. Finally, in terms of income level and employment status, only in the sample of women did we find that employed residents experienced lower overall green exposure and green exposure of average travel time than those who were unemployed. Increase in income had a positive connection with improving the green exposure of average travel time only in the sample of men.

To explore the enhanced or weakened effect of various travel and individual characteristics on male and female exposure to green space, we added the interaction effect model using gender and other factors for testing. The research results suggest that women have a higher level of exposure to green space when travel occurred on a weekday vs. on the weekend (Fig. 4.13). Furthermore, women have a higher level of green exposure in average travel time than men do during each trip on a weekday (Fig. 4.13a). However, taking the factor of driving into account enhanced a man's opportunity to be exposed to urban green space, since it had a more significant positive connection with increasing overall green exposure in comparison with women. This signified that the overall green exposure of men was higher than that of women after each driving trip (Fig. 4.11a). Taking public transportation played a significant role both in enhancing overall green exposure compared to women and in giving women the advantage over men in green exposure of average travel distance (Figs. 4.11a and 4.12a, respectively). Taking into account individual characteristics, the key findings that emerged indicate that age, household type, and having a driver's license can all enhance advantage over women in traveling through urban green space, multiplying the opportunities for men to be exposed to and utilize green spaces. First, the increase in age expanded the gender difference in overall green exposure: the older a man was, the higher the overall green exposure obtained for each trip than a woman of the same age (Fig. 4.11b).

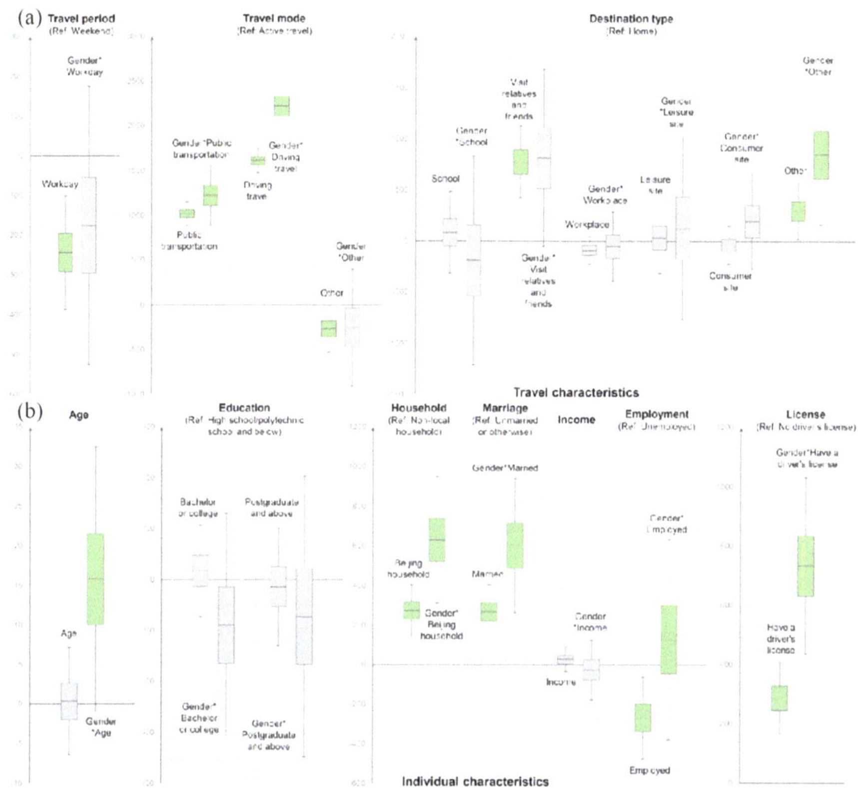

Fig. 4.11 Interaction effects of gender with (**a**) travel behavior and (**b**) individual attributes on overall green perception

Second, a positive correlation also existed between having a local household in Beijing and enabling men to experience higher levels of overall green exposure and green exposure of average travel time than women (Figs. 4.11b and 4.13b, respectively). Third, having a driver's license enhanced the overall green exposure and the green exposure of average travel time of men, however, it narrowed the gap between them in green exposure of average travel distance (Figs. 4.11b, 4.12b, and 4.13b, respectively). Furthermore, marital status was one of the key factors that intensified the gender difference, which was manifested in married men having a greater likelihood of increasing their overall green exposure for each trip (Fig. 4.11b). Being married also increased a man's green exposure of average travel distance (Fig. 4.12b). Last but not least, employment was a complex factor as well. Men who were employed had a higher overall green exposure than those who were unemployed, whereas women who were employed showed a lower level of overall green space exposure than those who were not employed (Fig. 4.11b).

The results showed that men generally have some advantages over women in terms of overall green exposure levels, while women were more likely to have a

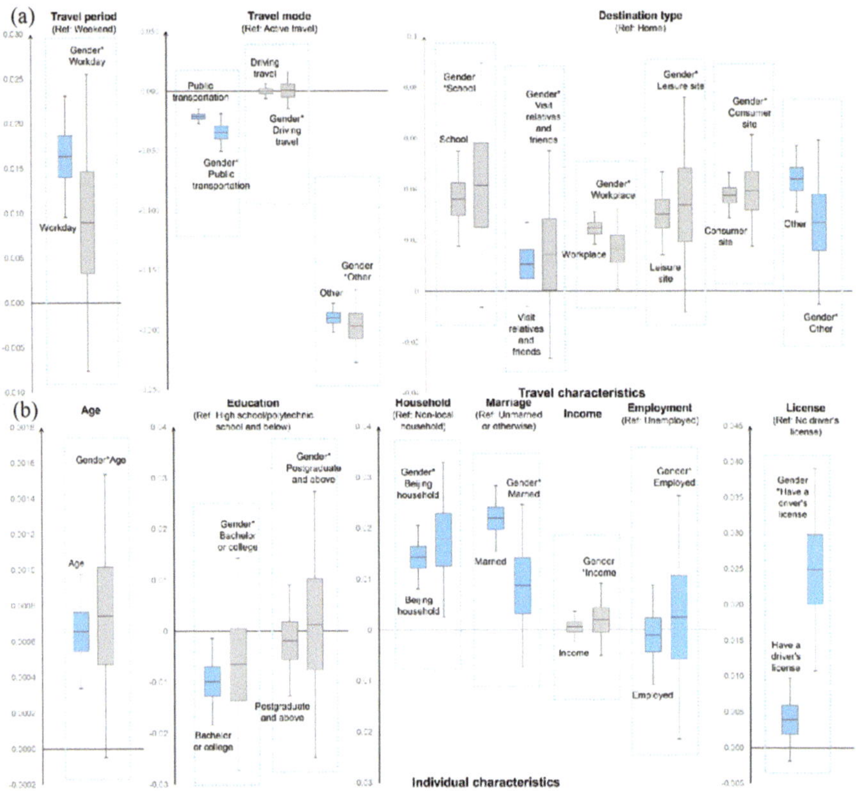

Fig. 4.12 Interactions between gender and (**a**) travel behavior and (**b**) personal factors in green perception per unit distance

higher level of green exposure in terms of travel distance. Since the range of women's daily mobility tends to be narrower, their travel distance is shorter, and because they assume more responsibilities for nonwork activities that compete for their time in daily life, the overall green exposure experienced by women during travel tends to be at lower levels than what men experience. The case is otherwise regarding our findings about green exposure of average travel distance. Women have higher levels of green exposure than men do. Since it eliminated the factor of distance, the green space quality (i.e., the average GVI) of the streets selected for travel became particularly important. Previous literature points out that women enjoy streetscape trees on their trips and show more willingness to make detours, which meant that they were more inclined to choose streets with high-quality green space. In addition, women tended to carry out more multifunctional travel and were more adept at synchronizing and coordinating complex travel routes than men were. Therefore, their daily travel could "bundle" various destinations and activities together. Green space planning along the routes chosen by women was often more extensive as well, and the green exposure of average travel distance they obtained during each trip

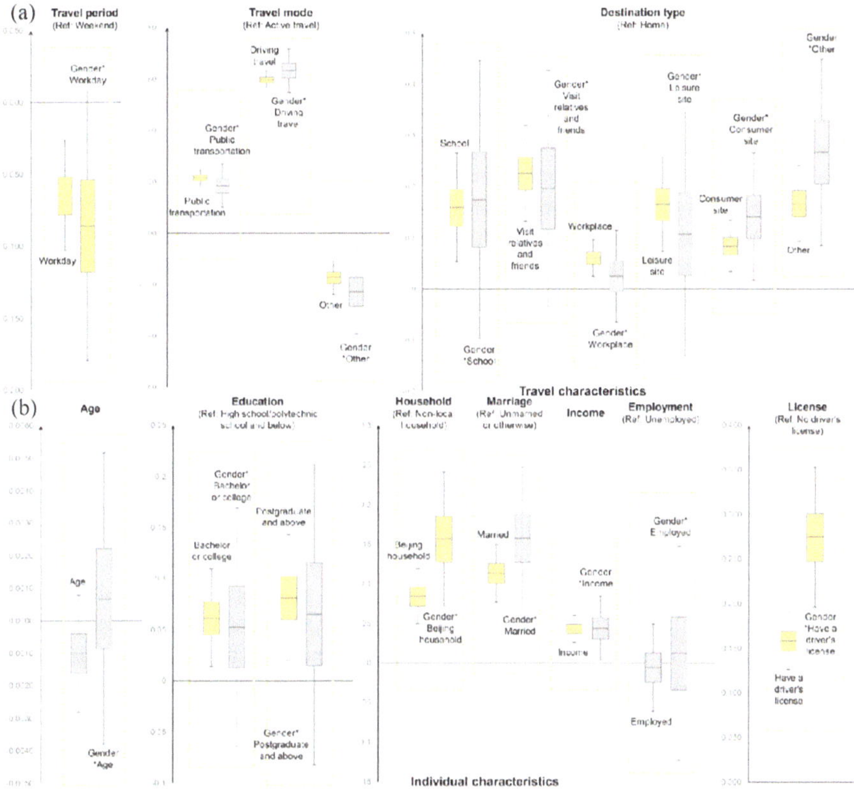

Fig. 4.13 Interaction effects of gender with (**a**) travel patterns and (**b**) individual traits on green perception per unit time

higher than that of men. However, as the number of cars in the family as well as the proportion of women holding driver's licenses increased, the frequency of trips by car among women also rose, the difference between men and women in travel speeds gradually decreased, and the green exposure of average travel time was no longer statistically significant.

Gender disparities in green exposure are associated with the travel and individual characteristics of residents, which is specifically reflected by different commuting conditions and safety awareness. First, some older studies indicate that women are more likely to find suitable occupations near home than men to achieve a work–life balance, although this contrast has declined over time. Coupled with the dual pressures of family and work responsibilities, the overall green exposure experienced by women on weekdays was far less than that on weekends, when they could travel more freely. For men, travel on weekdays tended to be concentrated in the morning and evening peaks. Hence, they often confronted traffic congestion, which affected travel speed and the ability to enjoy green space. The longer commuting distance of men meant more congestion on weekdays than it did for women. On weekends, when residents have relatively freer time for travel, they could choose when they would travel according to road conditions. Under such circumstances, the impact of traffic conditions on men's travel was significantly less than that on weekdays. Second, women's greater consideration of safety awareness is an important factor that affects their long-distance travel. Male drivers consistently overestimate their perceptual–motor skills and are more willing to drive longer distances than women are, while drivers tend to travel shorter distances but more frequently, so it is easier to be replaced with other travel modes, such as travel by foot or bike. Therefore, the positive correlation between having a driver's license and higher levels of green exposure is more obvious for men than for women, further strengthening their advantage over women in terms of overall green exposure.

In summary, gendered disparities in spatiotemporal constraints, behavioral preferences, and social roles contribute to green inequality in exposure levels. The green exposure of individuals across genders may be amplified or diminished by their travel patterns and personal attributes, necessitating integrated consideration of socioeconomic contexts rather than isolating individual factors. Research has shown that: (1) Residents showed significant gender disparities in their exposure to green space during travel, and men generally had slight advantages in terms of overall green exposure. (2) Women improved their green exposure through shorter distances and high-quality travel routes. (3) There were significant differences between men and women in the correlation between travel and individual characteristics, on the one hand, and urban green space exposure on the other, which could intensify or lessen the levels of such exposure by men and women, respectively.

At the level of practical policy implications, first, the service facilities and the space along streets around communities should be improved, especially in places with mixed functions, to provide more opportunities for women to be exposed to green space and thus obtain more of the benefits of green space from their surroundings. For some women, it is difficult to separate work and home, their working space tends to be not only a space for employment but also a space for living. Women also tend to travel in patterns that characterized by high green quality, shorter times, and closer distances. Therefore, improving the street landscape around communities and the quality of surrounding service facilities is advisable. Second, many residents regard daily travel as a way to supplement their green exposure to compensate for the shortcomings of the neighborhood environment. Therefore, vegetation along major streets in urban areas would improve residents' green exposure. Men tend to travel longer distances and use cars more often than women do, so increasing the amount and quality of green space along streets and highways with heavy traffic would increase the green exposure of men during travel even further. As for women, fewer have driver's licenses, drive cars less often, and travel more on public transportation than men do, so optimizing urban green space along public transportation routes would be a good approach to improve women's green exposure and reduce gender disparities in overall exposure to green space. Third, urban planners should take into account the needs and interests of women when building more green space in cities as well as encourage more women to participate in and influence decision-making about green space. If hypothesized differences exist between men and women in terms of physiology, psychology, ways of nurturing, and thinking styles, there must be differences in the aspects of their environment that men and women pay attention to. Policymakers should actively address the needs and interests of female users through gender-responsive interventions.

4.4 Urban Green Equity and Vegetation Chromatic Effects

Sections 4.1 and 4.2 have established that population mobility induces neighborhood effect biases, including the Neighborhood Effect Averaging Problem (NEAP) and Neighborhood Effect Polarization Problem (NEPP). Individuals experiencing environmental disadvantages may seek compensatory exposure through daily mobility or exhibit polarization due to convergence psychology, discrimination, or environmental segregation. Critically, however, NEAP and NEPP can transform under specific conditions. Even if NEPP is empirically observed in green exposure contexts, the coexistence of NEAP cannot be dismissed.

Revisiting the compensation hypothesis, humans possess fundamental environmental needs (Hall and Page 1999). When residential environments fail to meet these needs, individuals compensate through nonresidential exposure. Under this framework, residents in low-greenness areas might access higher-quality environments elsewhere. Yet two prerequisites must hold: Residents must subjectively perceive their local green exposure deficits, and they must actively pursue better environments. Current limitations in standardizing green space metrics (e.g., quantity/quality assessments) and quantifying fundamental demand for urban greenery undermine these prerequisites. Consequently, compensatory behaviors may fail to materialize. We thus posit that divergent green space metrics alter perceptions of exposure levels, ultimately impeding robust NEAP identification.

Indeed, urban green space contains a diverse range of plant landscapes and, therefore, cannot be accurately characterized solely by the degree of greenness, namely GVI (Wang et al. 2021). Considering that color of vegetation may play a critical role in attracting attention and eliciting landscape preferences, we supposed that the ornamental value of green space, which is closely related to the color, richness, and density of the vegetation, is also suitable for assessing the green exposure (Chen et al. 2018; Jørgensen and Anthopoulou 2007). Previous studies point out that vegetation color can bring stimulation and draw attention through visuals and arouse awareness and attention to the surrounding environment. The increase in the quantity and richness of vegetation color could contribute to the evaluation of the visual aesthetic quality of green spaces as well as enhance people's preference for green spaces (Lee and Hong 2013). Overall, the ornamental value of green space represented by vegetation color diversity is more likely to be perceived and incorporated as a target in residents' daily activities as a key green exposure feature. However, this key feature always seems to be overlooked in previous studies.

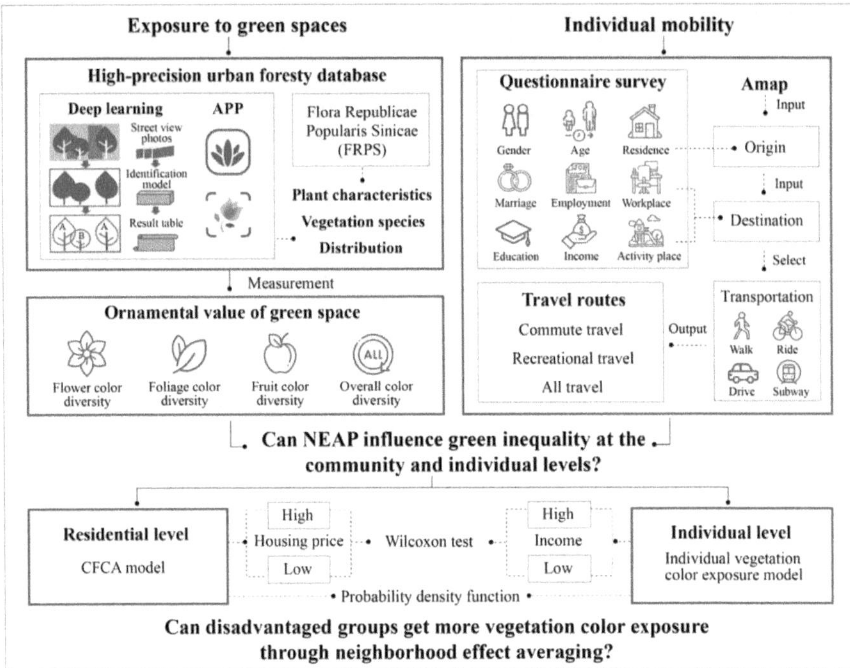

Fig. 4.14 Research framework for vegetation color inequality at individual scale

In this section, we will use the suburbs of Shanghai, China as an example. We established a high-precision urban vegetation database to measure the ornamental value of green space by leaf color, flower color, fruit color, and overall color diversity. Then, we evaluated vegetation color exposure separately at the community and individual levels, and examined green inequality issues related to the ornamental value of urban green space as well as potential compensatory mechanisms (Fig. 4.14).

4.4.1 *Individual-Level Measurement Methodology for Chromatic Exposure to Green Space Vegetation*

In order to conduct a thorough survey in Shanghai, we employed a multi-stage sample procedure due to the significant total number and diverse distribution of individuals residing in the suburbs and far suburbs of Shanghai. From April to July in 2017, we conducted a household questionnaire survey with a random sample of residents from 58 communities, taking into account the geographic location and regional development. The survey covered a variety of topics, including individual and household attributes, daily activity data, and activity diary data for two

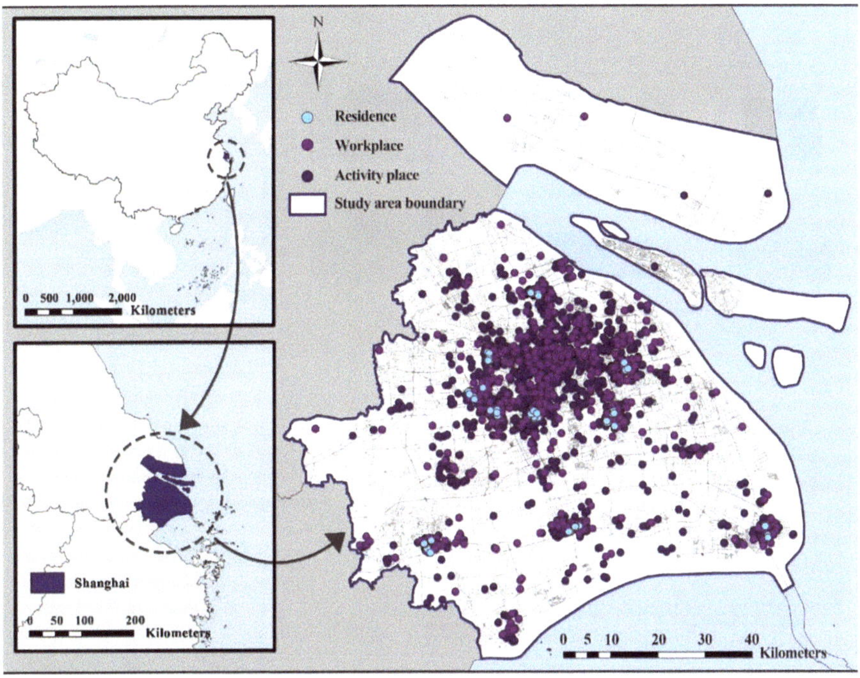

Fig. 4.15 Sampling areas and distribution map

consecutive days. A total of 1593 questionnaires were distributed, and 1140 valid responses were obtained, with a response rate of 71.56%. The data was cleaned using the following selection criteria after receiving a total of 1140 questionnaires: (1) residing in the current residential area for more than 10 years, and (2) having full information of all workplaces and activities, excluding those for the unemployed, the retirees, and minor children. Finally, we kept a reliable sample of 877 people from 54 different neighborhoods. See Fig. 4.15 for the distribution of respondents' residences, workplaces, and activity places. However, considering personal geoprivacy, all point elements are only used to indicate the relative geographical location, which cannot be located to the exact building or floor.

We have established a high-precision urban vegetation database (Fig. 4.16), which provided the possibility for the assessment of vegetation color diversity. Streetscape pictures and photo data from the plant identification app are the two main sources of vegetation data for this database. The former was crawled from Baidu Maps (https://map.baidu.com), covering four seasons from 2017 to 2021. We finally obtained about 350,000 streetscape photos with geographic coordinates, and recognized the plant information in each street scene image by deep learning algorithm. Green view index (GVI) could be calculated by the proportion of green pixels in the street pictures. The latter was derived from two apps, Xingse and aiPlants. For these two APPs' vegetation distribution records from 2017 to 2018, we employed a

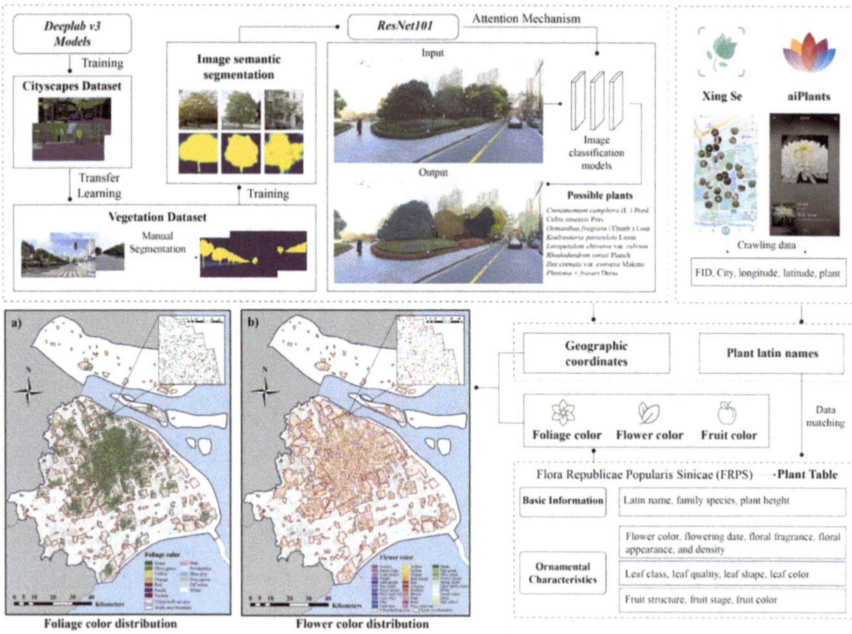

Fig. 4.16 High-precision vegetation database for urban green spaces. (**a**) Foliage color distribution, (**b**) flower color distribution

crawler to collect a total of 482,630 records, each of which provides the name and coordinates of the corresponding vegetation. Finally, we integrated the two datasets aforementioned and cleaned these datasets. The leaf color, flower color, and fruit color details for the corresponding vegetation were obtained from the Flora Reipublicae Popularis Sinicae (http://www.iplant.cn/frps). As a result, a detailed urban vegetation database spanning the area of Shanghai was created (5139 species, 3,246,555 vegetative plants). We can learn about the vegetation's spread and color performance in Shanghai through this database. However, plants vary in their timing of blossoming, yielding fruit, and shedding leaves. To mitigate the impact of seasonality, we assumed the highest richness in the year while calculating vegetation color diversity.

1. Vegetation Color Exposure at the Community Level

Based on the CFCA model framework, the key computational procedures encompass watershed delineation, green space attractiveness calculation, and supply–demand ratio calculation, all of which have been detailed in Sect. 4.1 of this chapter. For each community, all green spaces in the range are searched with the service range as the radius. The value of vegetation color exposure of each community is obtained by summing the supply–demand ratio of all selected green spaces (including the three travel modes):

A higher score of Vegetation Color Exposure indicates a higher level of vegetation color exposure in that community (Fig. 4.17).

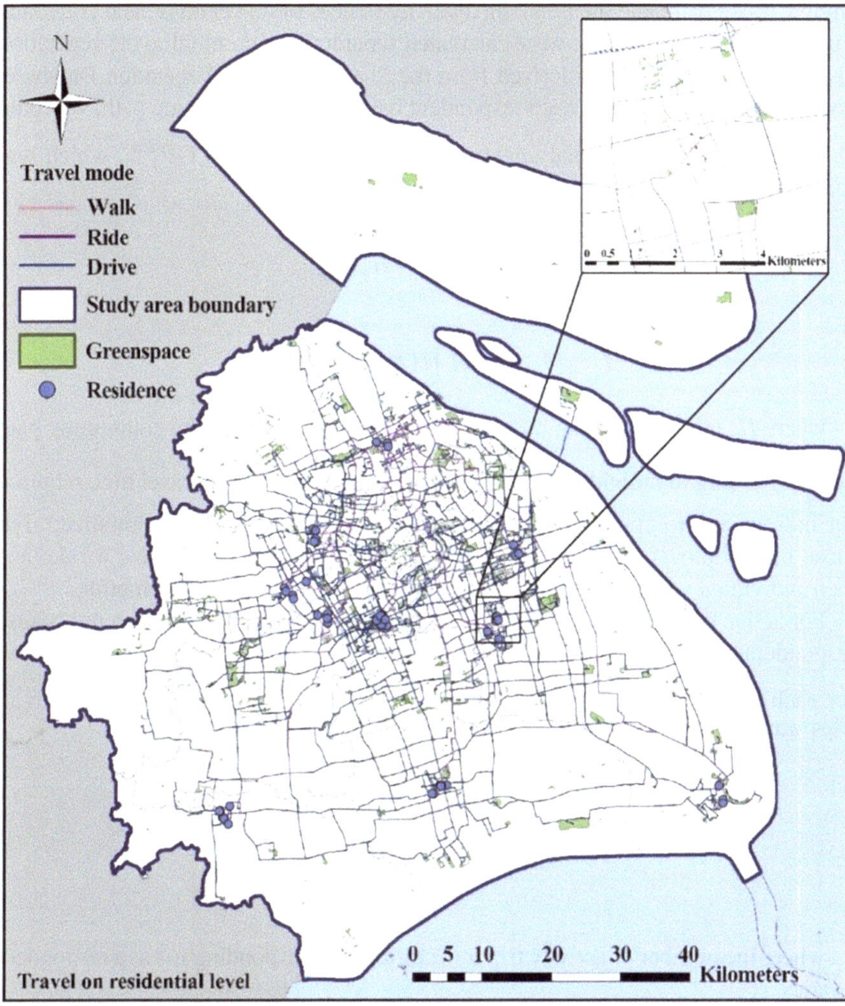

Fig. 4.17 Travel routes at community level

2. Vegetation Color Exposure at the Individual Level

We calculated individual vegetation color exposure for commuting trips, activity trips, and all trips based on the respondents' daily travel patterns (Wang et al. 2021). Given that commuting trips and activity trips are utilitarian in nature, we assume that individuals tend to opt for routes that minimize travel time. Therefore, by using the center of residential community as the starting point and the office or venue of daily activity as the destination, we selected the appropriate mode of transportation; we used Amap's shortest path planning function to determine the precise trip route. Additionally, given the limited likelihood of being exposed to vegetation color during subway travel, we disregarded the underground segments of the path and instead focused on the section between the origin and destination leading to the nearest subway station. A buffer zone with a radius of 200 m was constructed for each eligible travel path, and the Shannon diversity indices of flower color, leaf color, and fruit color in the buffer zone were calculated separately by combining the vegetation distribution and color data derived from the Shanghai Urban Vegetation Database. For commuting trips, since each respondent has a unique commuting path, we could directly obtain the individual vegetation color exposure $\text{Ind\_GE}_j^{\text{work}}$, which was calculated as follows:

$$\text{Ind\_GE}_j^{\text{work}} = H_{ij} \times \overline{T_{ij}},$$

$$\overline{T_{ij}} = \left(T_{ij} - \min T\right) / \left(\max T - \min T\right),$$

where H_{ij} is the vegetation color Shannon diversity index of the commuting path i corresponding to individual j within a 200 m buffer. $\overline{T_{ij}}$ is the travel time required for individual j to complete commuting path i, which has been normalized. For those non-employed individuals without fixed workplace coordinates, we defined their individual vegetation color exposure values to be 0 during commuting.

For active trips, the calculation of vegetation color exposure is different because respondents' multiple activity locations may generate various trajectories. Therefore, for each respondent j, their vegetation color exposure $\text{Ind\_GE}_j^{\text{act}}$ during activity trips was calculated as follows:

$$\text{Ind\_GE}_j^{\text{act}} = \frac{\sum_{i=1}^{n} H_{ij} \times \overline{T_{ij}}}{n},$$

where the number of activity trip trajectories n corresponding to each respondent j varies.

After completing the calculation of vegetation color exposure for each participant's commuting trips and activity trips, we integrated this as follows: since commuting trips occur almost every weekday, we assigned them higher weights; activity trips were assigned lower weights accordingly due to their random occurrence. In summary, the calculation formula of all trips' vegetation color exposure Ind_GE_j was:

$$\text{Ind\_GE}_j = \{ \begin{array}{ll} 5 \cdot \text{Ind\_GE}_j^{\text{work}} + 2 \cdot \text{Ind\_GE}_j^{\text{act}}, & \text{Ind\_GE}_j^{\text{work}} > 0 \\ 7 \cdot \text{Ind\_GE}_j^{\text{act}}, & \text{Ind\_GE}_j^{\text{work}} = 0 \end{array}.$$

For full-time employed people with fixed workplace coordinates and commuting paths, their individual-level vegetation color exposure was weighted based on the

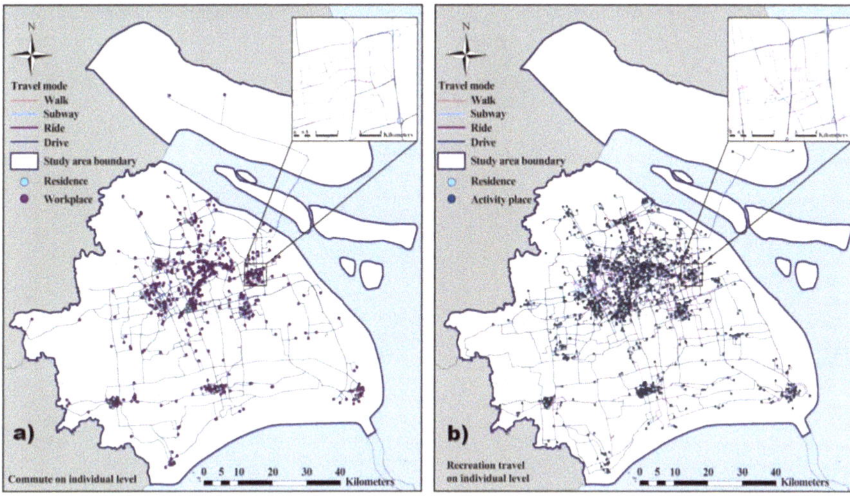

Fig. 4.18 Individual-level travel paths, including commuting travel path (**a**), and leisure travel path (**b**)

vegetation color exposure obtained during commuting and activity trips. For non-employed people without fixed workplace coordinates, their vegetation color exposure was evaluated solely based on the green landscape around their activity trip trajectories (Fig. 4.18).

3. Test for NEAP of Vegetation Color Exposure

We used the Wilcoxon test and probability density functions to examine whether NEAP exists in vegetation color exposure and whether it affects the green inequality at the community and individual levels. To be specific, we divided communities into two groups based on median housing price and divided the respondents into two groups according to their median incomes. By Wilcoxon test, we evaluated variations in exposure to vegetation color between different economic level groups at the community and individual levels, respectively. Next, we compared the results of the two levels to see whether green inequality was lessened by the compensatory behavior during individual mobility. Then, to further test the existence of NEAP, we plotted probability density functions of residence-based exposure and mobility-based exposure to assess whether GE^{I} tends to converge to the mean exposure value.

4.4.2 *Individual-Level Variations in Vegetation Chromatic Exposure and Their Underlying Mechanisms*

The results showed that the exposure to different types of vegetation varies significantly in terms of people's economic status. High-income areas often had greater exposure to vegetation color, with mean values for each indicator nearly twice as

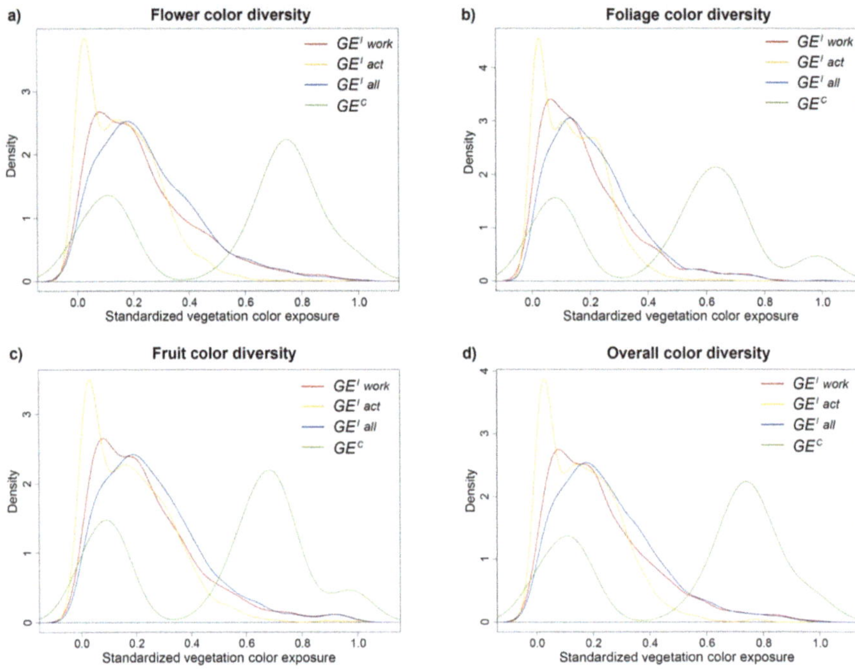

Fig. 4.19 Probability density curves of vegetation color exposure based on residence and mobility. (**a**) Flower color diversity, (**b**) foliage color diversity, (**c**) fruit color diversity, (**d**) overall color diversity

high as those in low-income areas. At the individual level, there is no discernible difference in vegetation color exposure between people when simply commuting is taken into account. When only activity visits are considered, the low-income group's daily activities expose them to much more vegetation color than those of the high-income group. Interestingly, when commuting and activity trips were taken into consideration simultaneously, the significant difference in vegetation color exposure between income groups vanished. Additionally, in the test that takes only activity travel account, all four indices of vegetation color richness showed significance. Among them, the mean value of comprehensive color richness is higher than the flower color richness, followed by the fruit color richness, and finally, the leaf color richness. In contrast to the results at the community level, however, the differences in vegetation color exposure between the high- and low-income groups were comparatively small. In addition, we compared the difference of GVI at community level and individual level. The results show that there is no significant difference in green exposure between communities of high- and low-economic level. Likewise, high- and low-income groups show no discernible difference in terms of perceived greenness during personal mobility. Figure 4.19 shows the probability density functions of residence-based exposure and mobility-based exposure for all samples, where is divided by commuting trips, activity trips, and all trips. It is clear that residence-based exposure tends to be polarized in all assessments of vegetation color exposure, whereas when we take daily mobility into account, more individuals' exposure levels tend to be averaged. This finding clearly supports the NEAP.

The findings of this study demonstrate that there is a green inequality at the local level, i.e., that communities with higher-income levels experience higher amounts of exposure to vegetation color. However, at the individual level, there is no significant difference in vegetation color exposure between high-income groups and low-income groups. It can be attributed to the neighborhood effect averaging problem (NEAP). The NEAP is anticipated to have an impact on vegetation color exposure, allowing people exposed to impoverished neighborhood surroundings to average their real exposure to vegetation color received through daily movement. There are several possible explanations. First, lower-income groups travel more frequently and use automobiles less frequently than higher-income groups. Longer travel time may have exposed passengers to greater vegetation color when the ornamental value of green space is close to the average level. Second, individuals are likely to compensate for their lack of exposure to vegetation color by altering the geographic context in where they live, such as selecting a workplace with a better environment, a route with higher vegetation coverage, or even a residential area with a nicer environment to obtain a higher level of vegetation color exposure. It is worth noting that high-income people tend to reside in communities with better economic conditions, and the rich green space resources near communities can substantially satisfy people's need for green space. Contrarily, low-income residents in communities with lower economic levels frequently struggle with the lack of nearby green spaces and the poor quality of those that do exist. As a result, they prefer to travel to far-off, high-quality green spaces for leisure activities and are further compensated for the exposure to vegetation color while doing so.

When GVI was used to represent green exposure level, no green inequality was found at both community level and individual level. However, when vegetation color diversity was adopted, the inequality occurred at the community level but not at the individual level. Instead, the level of vegetation color exposure at the individual level tended to average, which suggested the existence of NEAP. Our results seem to support previous speculation that different indicators of environmental exposure may affect the identification of NEAP. One possible explanation is that, compared with the degree of green, the disparity in ornamental value of green space (i.e., vegetation color difference) is easier to be perceived, the latter of which may result in individual compensatory behaviors, and thus NEAP occurs. In addition, vegetation color is beneficial for both physical and mental health. On the one hand, the landscape draws individuals toward it, prompting them to actively seek out and explore green spaces. As a result, they are able to experience a wide range of ecological benefits associated with such areas. On the other hand, specific plant types and vegetation colors have been proven to be more effective in relaxing the body, reducing anxiety, and improving mood. Therefore, vegetation color diversity is likely to be incorporated into residents' activities as a goal and become a key feature to promote individuals' daily mobility.

On the other hand, GVI can serve as a useful indicator of the quantity of green space, while vegetation color diversity is a more appropriate metric for assessing the quality of green space. In terms of spatial distribution, there might be some mismatch between the quantity and quality of green space. When it comes to quantity, numerous studies have discovered that communities with high economic levels typically have more greenery nearby, as evidenced by higher vegetation quantity and density, more green space per person, and better green space accessibility. In recent years, local governments in China have vigorously promoted the construction of green space; they built new parks and greenways to alleviate the problem of quantitative inequality. However, these measures alone may not ensure the fairness in the quality of green space. For instance, it was shown that communities with higher socioeconomic levels might have more green space but lower quality of green space in a study on the spatial distribution of green space in Shanghai. So, it is worth noting that there may be little difference in the GVI of current streetscape, but the diversity of vegetation color still exists, which indicates that the ornamental value of green space has the potential to be improved.

In summary, first, the disparity at the community level does exist, with economically privileged communities experiencing significantly higher levels of vegetation color exposure than economically underprivileged regions. Individually, however, there was no discernible difference in the vegetation color diversity that people with income gap were exposed to. Second, the ornamental value of green space might be a significant environmental feature associated with NEAP, as the variation in vegetation color richness is likely to be objectively observed and easily perceived by individuals. It is important to acknowledge that the survey sample contains a higher proportion of employed individuals with a higher level of education due to the limitations in respondents' participation. The sample bias may potentially affect the results. Despite persistent disparity at the community level, individual can obtain average exposure to vegetation color through daily movement. Secondly, we go into the environmental characteristics associated with NEAP. NEAP is usually related to environmental exposures that rely on daily mobility. As urban green space construction booming, green spaces are undoubtedly areas that people would pass through during the mobility. Given that the quantity and quality of green spaces are both factors affecting the daily movement, it seems that the occurrence of NEAP may be more strongly driven by the latter factor, which represented by the ornamental value of green space. We believe that this may be attributed to the following two reasons: Firstly, NEAP is usually explained by people's compensatory behaviors, precondition of which is the individual perception of the inferior exposure levels and their wishes to access to better environment. The vegetation color richness appears to be a crucial factor influencing people's perception compared to GVI. Secondly, there may be no considerable difference in the quantity of urban green space exposed to high- and low-income groups, but there is still inequality in the quality of green space. Therefore, from the perspective of urban decision makers, enhancing the quality of green space, such as the ornamental value of green space, is an important initiative to improve green inequality.

References

Bell K (2016) Bread and roses: a gender perspective on environmental justice and public health. Int J Environ Res Public Health 13(10):Article 10. https://doi.org/10.3390/ijerph13101005

Bolte G, Nanninga S, Dandolo L (2019) Sex/gender differences in the association between residential green space and self-rated health—A sex/gender-focused systematic review. Int J Environ Res Public Health 16(23):4818–4818. https://doi.org/10.3390/ijerph16234818

Bresnahan BW, Dickie M, Gerking SD (1997) Averting behavior and urban air pollution. Land Econ 73(3):340–340. https://doi.org/10.2307/3147172

Burgoine T, Monsivais P (2013) Characterising food environment exposure at home, at work, and along commuting journeys using data on adults in the UK. Int J Behav Nutr Phys Act 10(1):85–85. https://doi.org/10.1186/1479-5868-10-85

Chaix B, Méline J, Duncan S, Merrien C, Karusisi N, Perchoux C, Lewin A, Labadi K, Kestens Y (2013) GPS tracking in neighborhood and health studies: a step forward for environmental exposure assessment, a step backward for causal inference? Health Place 21:46–51. https://doi.org/10.1016/j.healthplace.2013.01.003

Chen M, Bargh JA (1999) Consequences of automatic evaluation: immediate behavioral predispositions to approach or avoid the stimulus. Personal Soc Psychol Bull 25(2):215–224

Chen X, Yang X (2014) Does food environment influence food choices? A geographical analysis through "tweets". Appl Geogr 51:82–89. https://doi.org/10.1016/j.apgeog.2014.04.003

Chen W, Albert C, von Haaren C (2018) The elderly in green spaces: exploring requirements and preferences concerning nature-based recreation. Sustain Cities Soc 38:582–593. https://doi.org/10.1016/j.scs.2018.01.023

Chidambaram B, Scheiner J (2021) Work-trip mode choice in Germany—affected by individual constraints or by partner interaction? Travel Behav Soc 24:231–244. https://doi.org/10.1016/j.tbs.2021.04.007

Crawford D, Timperio A, Giles-Corti B, Ball K, Hume C, Roberts R, Andrianopoulos N, Salmon J (2007) Do features of public open spaces vary according to neighbourhood socio-economic status? Health Place 14(4):889–893. https://doi.org/10.1016/j.healthplace.2007.11.002

Cutter W, Neidell M (2009) Voluntary information programs and environmental regulation: evidence from 'spare the air'. J Environ Econ Manag 58(3):253–265. https://doi.org/10.1016/j.jeem.2009.03.003

Ewing, R., and R. Cervero. 2010. Travel and the built environment: A meta-analysis. Journal of the American Planning Association 76(3):265–94.

Fagnani J (1987) Daily commuting time: the stakes for working mothers in France. Transport Res Rec J Transport Res Board 1135. https://trid.trb.org/view/282870

Fan CC, Sun M, Zheng S (2011) Migration and split households: a comparison of sole, couple, and family migrants in Beijing, China. Environ Plann A Econ Space 43(9):2164–2185. https://doi.org/10.1068/a44128

Hall CM, Page SJ (1999) The geography of tourism and recreation: environment, place and space. Routledge, London. https://uhra.herts.ac.uk/handle/2299/17776

Hall CM, Page SJ (2014) The geography of tourism and recreation: environment, place and space. Routledge, Abingdon

Hanson S, Hanson P (1980) Gender and urban activity patterns in Uppsala, Sweden. Geogr Rev 70(3):291–299. https://doi.org/10.2307/214257

Jones A, Hillsdon M, Coombes E (2009) Greenspace access, use, and physical activity: understanding the effects of area deprivation. Prev Med 49(6):500–505. https://doi.org/10.1016/j.ypmed.2009.10.012

Jørgensen A, Anthopoulou A (2007) Enjoyment and fear in urban woodlands—does age make a difference? Urban For Urban Green 6(4):267–278. https://doi.org/10.1016/j.ufug.2007.05.004

Kaczynski AT, Potwarka LR, Smale B, Havitz ME (2009) Association of parkland proximity with neighborhood and park-based physical activity: variations by gender and age. Leis Sci 31(2):174–191. https://doi.org/10.1080/01490400802686045

Kestens Y, Lebel A, Chaix B, Clary C, Daniel M, Pampalon R, Thériault M, Subramanian SV (2012) Association between activity space exposure to food establishments and individual risk of overweight. PLoS One 7(8):e41418–e41418. https://doi.org/10.1371/journal.pone.0041418
Kim J, Kwan M-P (2018) Beyond commuting: ignoring individuals' activity-travel patterns may lead to inaccurate assessments of their exposure to traffic congestion. Int J Environ Res Public Health 16(1):89. https://doi.org/10.3390/ijerph16010089
Kim J, Kwan M-P (2021a) Assessment of sociodemographic disparities in environmental exposure might be erroneous due to neighborhood effect averaging: implications for environmental inequality research. Environ Res 195:110519. https://doi.org/10.1016/j.envres.2020.110519
Kim J, Kwan M-P (2021b) How neighborhood effect averaging might affect assessment of individual exposures to air pollution: a study of ozone exposures in Los Angeles. Ann Am Assoc Geogr 111(1):121–140
Kimpton A, Corcoran J, Wickes R (2016) Greenspace and crime. J Res Crime Delinq 54(3):303–337. https://doi.org/10.1177/0022427816666309
Konrad K (2015) Mobiler Alltag im Wandel des Geschlechterverhältnisses. Springer, Wiesbaden
Krenn PJ, Oja P, Titze S (2014) Route choices of transport bicyclists: a comparison of actually used and shortest routes. Int J Behav Nutr Phys Act 11(1). https://doi.org/10.1186/1479-5868-11-31
Kwan M (2000) Gender differences in space-time constraints. Area 32(2):145–156. https://doi.org/10.1111/j.1475-4762.2000.tb00125.x
Kwan M (2009) From place-based to people-based exposure measures. Soc Sci Med 69(9):1311–1313. https://doi.org/10.1016/j.socscimed.2009.07.013
Kwan M-P (2012) The uncertain geographic context problem. Ann Assoc Am Geogr 102(5):958–968. https://doi.org/10.1080/00045608.2012.687349
Kwan M (2013) Beyond space (As we knew it): toward temporally integrated geographies of segregation, health, and accessibility. Ann Assoc Am Geogr 103(5):1078–1086. https://doi.org/10.1080/00045608.2013.792177
Kwan M-P (2022) Human mobility and the neighborhood effect averaging problem (NEAP). In: Li B, Shi X, Zhu A-X, Wang C, Lin H (eds) New thinking in GIScience. Springer, Singapore, pp 95–101. https://doi.org/10.1007/978-981-19-3816-0_11
Lee G, Hong I (2013) Measuring spatial accessibility in the context of spatial disparity between demand and supply of urban park service. Landsc Urban Plan 119:85–90. https://doi.org/10.1016/j.landurbplan.2013.07.001
Liu T, He G, Lau AKH (2017) Avoidance behavior against air pollution: evidence from online search indices for anti-PM2.5 masks and air filters in Chinese cities. Environ Econ Policy Stud 20(2):325–363. https://doi.org/10.1007/s10018-017-0196-3
Lu Y, Chen R, Chen B, Wu J (2024) Inclusive green environment for all? An investigation of spatial access equity of urban green space and associated socioeconomic drivers in China. Landsc Urban Plan 241:104926. https://doi.org/10.1016/j.landurbplan.2023.104926
Ma X, Li X, Kwan M, Chai Y (2020) Who could not avoid exposure to high levels of residence-based pollution by daily mobility? Evidence of air pollution exposure from the perspective of the neighborhood effect averaging problem (NEAP). Int J Environ Res Public Health 17(4):1223–1223. https://doi.org/10.3390/ijerph17041223
Markevych I, Schoierer J, Hartig T, Chudnovsky A, Hystad P, Dzhambov AM, de Vries S, Triguero-Mas M, Bräuer M, Nieuwenhuijsen M, Lupp G, Richardson E, Astell-Burt T, Dimitrova D, Feng X, Sadeh M, Standl M, Heinrich J, Fuertes E (2017) Exploring pathways linking greenspace to health: theoretical and methodological guidance. Environ Res 158:301–317. https://doi.org/10.1016/j.envres.2017.06.028
Matthews SA (2008) The salience of neighborhood. Am J Prev Med 34(3):257–259. https://doi.org/10.1016/j.amepre.2007.12.001
Morris S, Guell C, Pollard TM (2019) Group walking as a "lifeline": understanding the place of outdoor walking groups in women's lives. Soc Sci Med 238:112489–112489. https://doi.org/10.1016/j.socscimed.2019.112489

Nawrath M, Kowarik I, Fischer LK (2019) The influence of green streets on cycling behavior in European cities. Landsc Urban Plan 190:103598–103598. https://doi.org/10.1016/j.landurbplan.2019.103598

Park YM, Kwan M (2016) Individual exposure estimates may be erroneous when spatiotemporal variability of air pollution and human mobility are ignored. Health Place 43:85–94. https://doi.org/10.1016/j.healthplace.2016.10.002

Perchoux C, Chaix B, Brondeel R, Kestens Y (2016) Residential buffer, perceived neighborhood, and individual activity space: new refinements in the definition of exposure areas—the RECORD Cohort Study. Health Place 40:116–122. https://doi.org/10.1016/j.healthplace.2016.05.004

Sarkar C, Webster C, Pryor M, Tang D, Melbourne SJ, Zhang X, Liu J (2015) Exploring associations between urban green, street design and walking: results from the greater London boroughs. Landsc Urban Plan 143:112–125. https://doi.org/10.1016/j.landurbplan.2015.06.013

Scheiner J (2016) Time use and the life course: a study of key events in the lives of men and women using panel data. Eur J Transp Infrastruct Res 16(4):Article 4. https://doi.org/10.18757/ejtir.2016.16.4.3163

Scheiner J, Holz-Rau C (2012) Gendered travel mode choice: a focus on car deficient households. J Transp Geogr 24:250–261. https://doi.org/10.1016/j.jtrangeo.2012.02.011

Schipperijn J, Ekholm O, Stigsdotter UK, Toftager M, Bentsen P, Kamper-Jørgensen F, Randrup TB (2010) Factors influencing the use of green space: results from a Danish national representative survey. Landsc Urban Plan 95(3):130–137. https://doi.org/10.1016/j.landurbplan.2009.12.010

Schwanen T, Kwan M, Ren F (2008) How fixed is fixed? Gendered rigidity of space–time constraints and geographies of everyday activities. Geoforum 39(6):2109–2121. https://doi.org/10.1016/j.geoforum.2008.09.002

Shafran-Nathan R, Yuval, Levy I, Broday DM (2016) Exposure estimation errors to nitrogen oxides on a population scale due to daytime activity away from home. Sci Total Environ 580:1401–1409. https://doi.org/10.1016/j.scitotenv.2016.12.105

Silm S, Ahas R, Nuga M (2013) Gender differences in space—time mobility patterns in a postcommunist city: a case study based on mobile positioning in the suburbs of Tallinn. Environ Plann B Plann Des 40(5):814–828

Smith LN (2018) A disciplined approach to neural network hyper-parameters: part 1—learning rate, batch size, momentum, and weight decay. Arxiv (Cornell Univ). https://doi.org/10.48550/arxiv.1803.09820

Solarz AK (1960) Latency of instrumental responses as a function of compatibility with the meaning of eliciting verbal signs. J Exp Psychol 59(4):239

Sreetheran M, van den Bosch CCK (2014) A socio-ecological exploration of fear of crime in urban green spaces—a systematic review. Urban For Urban Green 13(1):1–18. https://doi.org/10.1016/j.ufug.2013.11.006

Ta N, Zhao Y, Chai Y (2016) Built environment, peak hours and route choice efficiency: an investigation of commuting efficiency using GPS data. J Transp Geogr 57:161–170. https://doi.org/10.1016/j.jtrangeo.2016.10.005

Tan Y, Kwan M, Chen Z (2020) Examining ethnic exposure through the perspective of the neighborhood effect averaging problem: a case study of Xining, China. Int J Environ Res Public Health 17(8):2872–2872. https://doi.org/10.3390/ijerph17082872

Ulrich RS (1986) Human responses to vegetation and landscapes. Landsc Urban Plan 13:29–44. https://doi.org/10.1016/0169-2046(86)90005-8

Vedel SE, Jacobsen JB, Skov-Petersen H (2017) Bicyclists' preferences for route characteristics and crowding in Copenhagen—a choice experiment study of commuters. Transp Res A Policy Pract 100:53–64. https://doi.org/10.1016/j.tra.2017.04.006

Wang B, Xu T, Gao H, Ta N, Chai Y, Wu J (2021) Can daily mobility alleviate green inequality from living and working environments? Landsc Urban Plan 214:104179. https://doi.org/10.1016/j.landurbplan.2021.104179

Wu J, Wang B, Ta N, Zhou K, Chai Y (2020) Does street greenery always promote active travel? Evidence from Beijing. Urban For Urban Green 56:126886. https://doi.org/10.1016/j.ufug.2020.126886

Xu T, Wang S, Liu Q, Kim J, Zhang J, Ren Y, Ta N, Wang X, Wu J (2023) Vegetation color exposure differences at the community and individual levels: an explanatory framework based on the neighborhood effect averaging problem. Urban For Urban Green 86:128001. https://doi.org/10.1016/j.ufug.2023.128001

Yu H, Deng Y, Xu S (2018) Evolutionary pattern and effect of administrative division adjustment during urbanization of China: empirical analysis on multiple scales. Chin Geogr Sci 28(5):758–772. https://doi.org/10.1007/s11769-018-0990-2

Zenk SN, Schulz AJ, Matthews SA, Odoms-Young A, Wilbur JE, Wegrzyn LR, Gibbs KP, Braunschweig C, Stokes C (2011) Activity space environment and dietary and physical activity behaviors: a pilot study. Health Place 17(5):1150–1161. https://doi.org/10.1016/j.healthplace.2011.05.001

Zhang H, Chen B, Sun Z, Bao Z (2012) Landscape perception and recreation needs in urban green space in Fuyang, Hangzhou, China. Urban For Urban Green 12(1):44–52. https://doi.org/10.1016/j.ufug.2012.11.001

Zivin JG, Neidell M (2009) Days of haze: environmental information disclosure and intertemporal avoidance behavior. J Environ Econ Manag 58(2):119–128. https://doi.org/10.1016/j.jeem.2009.03.001

Chapter 5
Future Directions in Urban Green Space Equity Research

5.1 Future Research Directions Across Three Scales

1. Global Scale
Recent studies offer exemplary frameworks and perspectives; yet more evidence is needed to inform urban planning worldwide. Chinese research shows that 72% of rapidly urbanizing cities have improved the equity of their green space provision—an encouraging sign for addressing global green space inequality. Several factors may explain this trend: (1) Governance and land policy. China's municipal management system gives government a dominant role in integrated planning, and the land-transfer system stipulates that green space use rights are non-transferable. Government thus retains the authority to requisition land for public green space construction, securing a more equitable distribution of resources. (2) Built-environment characteristics. Chinese cities typically exhibit compact, intensive land use, which helps equalize the spatial distribution of accessible green space. Compact urban form and a dense public-transport network also encourage a higher share of green travel, increasing daily contact with greenery through routine mobility. These unique policy frameworks, planning norms, and spatial environments distinguish Chinese cities from their Western counterparts. Investigating whether the patterns and drivers of green space inequality in China differ from those in the West can illuminate equity dynamics in developing countries, uncover policy, environmental, and behavioral mechanisms, and reduce the geographic imbalance that currently characterizes international research on green space justice.

2. City Scale
Models of green space accessibility need further refinement to capture the dynamics created by population mobility, traffic congestion, and seasonal phenology. Composite indicators of green space quality should also become more comprehensive. Because parks differ in size, available recreational activities, and users' actual or perceived experiences, uniform standards devised in earlier studies may be

J. Wu, *Urban Green Space Equity in China*, Urban Sustainability,
https://doi.org/10.1007/978-981-95-5771-4_5

ill-suited to every site. Future work must also minimize the impact of temporal mismatches among datasets, test findings in a wider range of cities, and investigate equity dimensions that go beyond household income.

3. Individual Scale
Understanding remains limited regarding under what conditions the neighborhood effect of urban green space operates as an equalizing force and under what conditions it becomes polarizing. Future work must extend the measurement dimensions, temporal frameworks, personal attributes, and regional contexts used to evaluate green space. It is essential to examine how socioeconomic status, acting through distributional, procedural, and interactional justice, shapes individuals' environmental exposure and mobility, and to explore whether restoring the effectiveness of NEAP can mitigate environmental inequality. Accordingly, research should be broadened to more diverse scenarios and should probe, at the individual level, the mechanisms by which NEAP and NEPP arise or transform, thereby identifying strategies to forestall more severe environmental justice issues.

5.2 Viable Interdisciplinary Directions for Future Work

Beyond equity, research on urban green space spans health, ecology, and related domains. These strands are not isolated; they increasingly intersect and converge. Interdisciplinary integration will therefore be a key avenue for breakthroughs.

1. Health × Equity
This cross-field agenda should foreground the interactions and mediating pathways between social equity and health equity, and unpack how green space attributes differentially shape the formation of social capital. It is especially important to analyze how spatial heterogeneity—along urbanization gradients—moderates health benefits. Research needs to move beyond static exposure metrics and develop composite evaluation systems that integrate dynamic spatiotemporal behavior and subjective perception (e.g., intelligent park-environment assessment systems). Such tools can reveal the non-linear relationships among green space accessibility, spatial distribution, and health outcomes under conditions of high population density and the expansion of informal settlements. In parallel, longitudinal cohort designs and intervention experiments are needed to trace the bidirectional effects of greening interventions on social and environmental equity, and to observe, over time, the co-evolution of green space use patterns and social networks—while attending to potential harms arising from management deficits. Study designs should strengthen transparency in causal inference, clarifying how population mobility, green space attrition, and health inequality interact amid rapid urbanization, and quantifying the dispossessing effects of land-use competition (e.g., housing expansion crowding out green space) on the health rights of vulnerable groups. At the same time, fine-grained equity analyses—via subgroup analyses—should reveal how power distributions, capital flows, and spatial justice shape the heterogeneous distribution of

green space benefits across income, age, and ethnic groups, remaining alert to imbalances in spatial rights that may hide beneath the surface of "equal coverage." For example, a surface greenway may boost vitality in middle-class neighborhoods yet, by adding impervious pavement, increase pluvial flood risk in low-income communities—tradeoffs that call for multi-scale spatial simulation and participatory appraisal.

2. Equity × Ecology
This agenda will deepen the theoretical and practical linkages between other realms of environmental justice—such as climate justice—and green justice, while expanding empirical cases for different subpopulations and cities in the Global South, and exploring differentiated spatial needs for women and people with disabilities. Current evidence shows that the absence of a cross-sector justice perspective hampers the systematic identification of intersectional exclusion mechanisms—across gender, race, and class—in urban green space planning. Moreover, the equitable benefits of blue–green infrastructure in mitigating heat islands and regulating microclimates have yet to be tightly connected to core environmental justice concerns such as community economic vulnerability and racialized spatial deprivation. Meeting these challenges requires ecological classification schemes that go beyond traditional green space typologies to include functional indicators—hydrological connectivity, the heat-regulation efficiency of tree canopies, and related metrics. It also calls for cross-scale assessment tools for green justice that integrate micro-level mechanisms at the community scale (e.g., green coverage relative to income gaps) with macro-level dynamics at the regional scale (e.g., gentrification induced by park siting), ultimately generating theory and evidence that balance ecological performance with social equity.

3. Equity × Health × Ecology
In practice, the overarching goal is a "triple win." On the measurement front, research should shift from single-endpoint outcomes to a complex-systems perspective, building life-cycle cost–benefit models and, in particular, filling the gap in long-term evaluations that move beyond cross-sectional evidence. Although urban greening interventions show promise, most studies lack follow-up beyond 12 months, making it difficult for policymakers to judge sustainability and potential attenuation of benefits. Long time-series designs are thus needed to elucidate behavioral adaptation, cumulative ecological-function effects, and the restructuring of social networks. Study designs should also attend to the diversity of interventions—green walls, community gardens, urban agriculture—and their heterogeneous impacts on health, well-being, and social equity, while systematically assessing possible adverse cascades, such as gentrification that intensifies exclusion or vegetation layouts that impede local air flow. Addressing these complexities requires interdisciplinary, multidimensional monitoring that simultaneously captures environmental, health, social, and economic indicators; probes the deep linkages among power structures, capital flows, and spatial justice; and provides the theoretical basis for elevating green space interventions from ecological engineering to instruments of social governance. Finally, government–academia–community co-production

and participatory design should strengthen the spatial voice of marginalized groups, advancing the triple goals of health promotion, environmental justice, and social inclusion.

Despite the remaining challenges, the innovations and exploratory studies accomplished to date are highly valuable. By confronting the limitations of past work, experimenting with diverse models, and embracing interdisciplinary analytical methods, researchers have enriched the study of urban green space equity and offered concrete recommendations for improving green space governance in China.

Index

J. Wu, *Urban Green Space Equity in China*, Urban Sustainability,
https://doi.org/10.1007/978-981-95-5771-4

MIX
Papier aus verantwortungsvollen Quellen
Paper from responsible sources
FSC® C105338

If you have any concerns about our products,
you can contact us on
ProductSafety@springernature.com

In case Publisher is established outside the EU,
the EU authorized representative is:
Springer Nature Customer Service Center GmbH
Europaplatz 3, 69115 Heidelberg, Germany

Printed by Libri Plureos GmbH
in Hamburg, Germany